Practical Astro

Springer
*London
Berlin
Heidelberg
New York
Barcelona
Budapest
Hong Kong
Milan
Paris
Santa Clara
Singapore
Tokyo*

Other titles in this series

The Modern Amateur Astronomer
Patrick Moore (Ed.)

The Observational Amateur Astronomer
Patrick Moore (Ed.)

Telescopes and Techniques
C.R. Kitchin

Small Astronomical Observatories
Patrick Moore (Ed.)

The Art and Science of CCD Astronomy

David Ratledge (Ed.)

Springer

David Ratledge, BSc, CEng, MICE
24 The Common, Adlington, Chorley,
Lancashire PR7 4DR, UK
Email: 100632.2746@compuserve.com

Cover illustration: The four insets show: Saturn by Gregory Terrance; Galaxy NGC253 by Nik Szymanek and Ian King; Omega Centauri by Tim Puckett; Horsehead Nebula by David Ratledge.

ISBN 3-540-76103-9 Springer-Verlag Berlin Heidelberg New York

British Library Cataloguing in Publication Data
The art and science of CCD astronomy
   1. Astronomy 2. Astronomy – Observations 3. CCD cameras
   I. Ratledge, David
   522.6
ISBN 3540761039

Library of Congress Cataloging-in-Publication Data
The art and science of CCD astronomy / David Ratledge, (ed.)
   p. cm. – (Practical astronomy)
   Includes bibliographical references and index.
   ISBN 3-540-76103-9 (pbk.: alk. paper)
   1. Astronomical instruments–Congresses. 2. Charge coupled devices–Congresses. I. Ratledge, David, 1945- . II. Series.
QB84.5.A78   1996   96-43811
522′.2-dc20   CIP

Apart from any fair dealing for the purposes of research or private study, or criticism or review, as permitted under the Copyright, Designs and Patents Act 1988, this publication may only be reproduced, stored or transmitted, in any form or by any means, with the prior permission in writing of the publishers, or in the case of reprographic reproduction in accordance with the terms of licences issued by the Copyright Licensing Agency. Enquiries concerning reproduction outside those terms should be sent to the publishers.

© Springer-Verlag London Limited 1997
Printed in Great Britain

The use of registered names, trademarks, etc. in this publication does not imply, even in the absence of a specific statement, that such names are exempt from the relevant laws and regulations and therefore free for general use.

The publisher makes no representation, express or implied, with regard to the accuracy of the information contained in this book and cannot accept any legal responsibility or liability for any errors and omissions that may be made.

Typeset by EXPO Holdings, Malaysia
Printed by Cambridge University Press
58/3830-543210   Printed on acid free paper

Preface

The computer beeps in a darkened observatory. A message appears: "transferring..." In a few impatient seconds the image will be safely downloaded. A click with the mouse on *Visualise* and behold, it's there! The Horsehead Nebula, which is totally invisible to the eye through the telescope, appears sharp and clear on the computer screen. All over the world this is repeated nightly in amateur observatories, be they in urban backgardens or under darker skies out of town. The digital revolution has begun – the charge-coupled device, the CCD, has arrived.

However, that revolution comes at a price – it brings with it its own new problems, ones we have not had to solve before. How do you focus a CCD camera? How do you centre an object on it? Can you find guide stars? What are flat fields? Now the good news – these and many other problems that the beginner CCD user will meet are being solved every night. The main part of this book describes the techniques and solutions of twelve amateurs with a combined experience in imaging of over forty years. These amateurs are producing images rivalling those of professional observatories of only tens of years ago. By studying their methodology, the reader will be able to short-cut the learning curve to successful CCD imaging and progress painlessly to that level, and even beyond. Inevitably, new and better CCDs will appear, but the techniques and examples of good practice outlined here will still be applicable for many years to come. The results will just get better!

The Origin of This Book

Not only are CCDs a product of the digital age, but so is this book. All the contributors publish their images on the Internet and regularly learn from, and share tips with, one another. It was this dialogue which revealed that a book such as this is needed, where such experience can be assembled and more widely shared. Without the Internet, the bringing together of experienced CCD users from the USA, Canada, the UK and Belgium would have been a nightmare. So it is no accident that all the contributors have Internet email addresses. The Internet has enabled daily contact to be kept and the contributions to be discussed and merged into this book. All contributors have agreed to their Internet addresses being published, so if the reader requires further information from them they are only a simple email away.

The Scope

Virtually every aspect of current CCD imaging and problems that the beginner is likely to face has been covered by our contributors. The contributors themselves have been chosen primarily for the quality of their work over the years, and although they have concentrated on one subject it should be pointed out that they are equally proficient in most areas of CCD imaging. They were also selected with a wide variety of equipment, locations and budgets in mind, so that an extensive range of techniques and solutions can be compared. Not everyone will have an unlimited budget or the luxury of dark skies, but whether you have these or not there is much practical advice available. It also should be made clear that they are amateur astronomers, and as such their opinions are free from any commercial bias (with one exception explained in the text) that could perhaps creep in from manufacturers or resellers. In other words their opinions and findings can be taken at face value and are the more useful for that.

Preface

After my general introduction to CCDs, our first specialist subject is lunar imaging. Dave Petherick from Canada explains how to tackle a subject which perhaps was considered until recently more suitable for film than silicon. Dave has made not only his telescope and his mirror but also his Cookbook CCD camera as well. His mosaic technique allows areas larger than the chip to be covered, and in stunning sharpness. His lunar CCD images have been featured in *Astronomy* magazine (his full Moon mosaic won $100 as Best Astrophoto of the Month in February 1994), *CCD Astronomy* and Canada's *Sky News* magazine. He has also made presentations at the Huronia Star Party and Starfest in southern Ontario.

Another subject suitable for CCD cameras that might surprise many is solar imaging. Brian Colville, a member of the same astronomical society as Dave, deals with this tricky but rewarding topic. Here we are using the CCD's digital ability rather than its light sensitivity. Although primarily concentrating on solar imaging, Brian covers many of the basics of CCD imaging which are equally applicable to other subjects. This talented Canadian duo can also combine their efforts, as was the case in their spectacular mosaic of Comet Hyakutake's tail (see the Colour Gallery, p. 16).

Gregory Terrance's images of the planets are some of the best taken by an amateur. This USA-based amateur is a renowned astrophotographer turned CCD imager. He explains the secrets of this rewarding subject, which extend not just to using the best camera and telescope techniques but also to employing the best location for seeing as well. Gregory's talents are not restricted solely to imaging; he has recently completed a CCD camera of his own manufacture using the Kodak KAF0400 chip.

The subject of comets has never been more popular, and fully justifies two chapters. This is one subject where the imager might not get a second chance, so it has to be right first time. Fortunately our two contributors provide the know-how to do just that. First Tim Puckett from Georgia, with over eighteen years' experience in photographing and imaging comets, explains where to find the necessary information, and how and where to report discoveries. He also explains his state-of-the-art methods for monitoring these objects and gives us a tantalising look into the future with remote automated imaging. Over on this side of the Atlantic, David Strange has produced a stunning catalogue of recent comets from his observatory in

Dorset, England. David also uses mosaics to good advantage, particularly when a prominent cometary tail is present. He is another whose work regularly appears in the astronomical press in both the USA and UK.

Deep-sky imaging is covered by Luc Vanhoeck from Belgium. Luc is well known to readers of *CCD Astronomy*, having carried out the review for that magazine of the popular SBIG ST8 CCD camera. He has also edited, for five years, *Heedal*, the magazine of the Flemish Astronomical Association. His nebula and galaxy images are all in remarkable detail and have been taken both "at home" and on holiday where the trials and tribulations of imaging away from base are amusingly recounted.

Wide-field imaging is a topic too little discussed or appreciated, despite being probably the easiest (and cheapest) way to get started in CCD imaging. John Sanford from Orange County Astronomers in the USA puts the record straight and explains the pros and cons of this subject, together with which lenses and filters are needed. John has an asteroid named after him (No. 5736) and he is an author in his own right, his book *Observing the Constellations* having been published in 1989.

Light pollution is probably going to get even worse before – we may hope – it gets better, and is probably the biggest problem facing the average amateur astronomer today. Adrian Catterall, who has successfully imaged for several years from within five miles of central London, has had to overcome some of the worst-polluted skies in the world. He explains the techniques which have enabled him to image objects that would be impossible visually or with film. At such a light-polluted location, Adrian relies on filters and has advice based on experience as to their best use. I too, living on the northern fringe of Greater Manchester and therefore imaging all the best objects through its orange glow, have had to adopt a methodology for imaging the invisible. I also cover some of the devices and gadgets which make CCD imaging that bit easier and, as a result, more enjoyable.

The double team of Nik Szymanek and Ian King are well known around the world for their tri-colour imaging, which they have refined over several years to the point where it now rivals the finest colour film images. These two, although UK-based, do not restrict themselves to their own country, making regular trips

Preface

with their CCD equipment to the dark skies of La Palma. They explain the intricacies of the tri-colour technique, which is probably the hardest facet of CCD imaging to master.

Finally, George Sallit from Reading, England, covers a range of subjects such as asteroid discovery (remember the popular press (mis)reports of his discovery of a new minor planet?), astrometry and photometry. George shows that there is more to imaging than just pretty pictures, delightful though they may be.

The Colour Gallery allows the contributors to show their versatility with images covering a range of objects rather than the mere one they were asked concentrate on in their own chapter. There is some duplication of content. This is deliberate and has been done so that different cameras, telescopes, focal lengths, locations and, of course, techniques can be compared.

Acknowledgements

First and foremost I must thank our contributors who (usually) responded enthusiastically to my suggestions.

My thanks must also go to the proof-readers, my wife Julie, Peter Cleland and Pat McKenna. Needless to say, any mistakes still present are not their responsibility.

Finally I am grateful to John Watson, Managing Director at Springer UK, for fully supporting and backing the idea of a CCD book on the combined experiences of twelve amateurs from around the world.

Lancashire,   David Ratledge
UK.

Contents

1. **An Introduction to CCDs**
   David Ratledge 1

2. **Lunar Imaging with the Cookbook CCD Camera**
   Dave Petherick 19

3. **Solar CCD Imaging**
   Brian Colville 33

4. **High-Resolution Planetary Imaging**
   Gregory A. Terrance 49

5. **The Comet Watch Program**
   Tim Puckett 61

6. **Imaging Comets**
   David Strange 73

7. **Nebulae and Galaxies in High Resolution**
   Luc Vanhoeck 81

8. **Wide-Field Imaging**
   John Sanford 93

9. **CCD Imaging from the City**
   Adrian Catterall 101

10. **Overcoming Light Pollution**
    David Ratledge 111

11. **Tri-Colour CCD Imaging**
    Nik Szymanek and Ian King 123

12. **Beyond Pretty Pictures**
    George Sallit 135

Appendix A	CCD Camera Manufacturers	145
Appendix B	Sizes of Some Common CCD Chips	147
Appendix C	CCD Field of View for Various Focal Lengths	149
Appendix D	Sizes of Some Deep-Sky Objects	151
Appendix E	Bibliography	153
Appendix F	Astronomical Image Processing Software	155
Appendix G	Glossary of Terms	157

Contributors 161

Followed by

Colour Gallery: Some of the best images from the Contributors

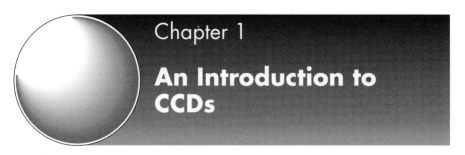

Chapter 1
An Introduction to CCDs

David Ratledge

Background

The charge-coupled device, or CCD, was developed at Bell Laboratories over twenty-five years ago, initially as a storage device, but what has revolutionised astronomy is its use as an optical detector. The revolution is as big as those made by the invention of the telescope and photography in the past. Like photography, CCDs have the ability to record objects too faint for the eye to see, by exposing them for many minutes, even hours. Photographic emulsions have evolved tremendously, but they still record only around 3 to 5% of the light that lands on them, even when hypersensitised. On the other hand, CCDs record around 30–50%, and even more in professional systems. That is a gain of ten times in efficiency! Better still, they do not suffer from low-light reciprocity failure, the bane of astrophotography, where the film fails to live up to its speed rating and effectively "slows down" because of the low light levels of the object being photographed. A telescope equipped with a CCD camera suddenly behaves like one with ten times the area. In other words, a 20 cm (8 in) telescope becomes as sensitive as a 60 cm (24 in) one! There is no amateur astronomer alive who does not dream of a bigger telescope, and with a CCD camera that dream becomes a reality. Viewed in this way the perceived high cost of CCD cameras is but an illusion – they are in fact a bargain!

The Sky's the Limit!

When we equip our own telescope with a CCD, the number of objects that come within its grasp suddenly expands. No longer are we limited to the Messier Objects – they become almost too easy. The whole of the *New General Catalogue* (*NGC*), with its eight thousand faint galaxies, nebulae and clusters, comes into the range of the amateur astronomer, even one who has abandoned astrophotography because of light pollution. In addition, galaxies too faint for the NGC will be recorded, quasars too. The CCD provides a lifetime of imaging to look forward to.

What is a CCD?

The appearance of a CCD camera is at first sight off-putting, with its black fins and trailing wires, and with no viewfinder to look through. The actual CCD is the silicon chip buried inside under the protection of an optical window. The rest of the camera comprises electronics to record and digitise the signal, plus a cooling system to keep it cold. If you look closely at the CCD chip you will see on it a rectangle or square, which is the light-sensitive area. Under a microscope, this area would be revealed as being divided up like a tiny chessboard, into minute squares (Figure 1.1).

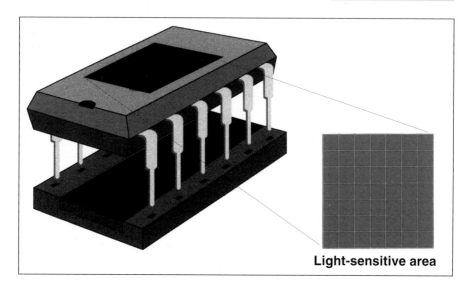

Figure 1.1 CCD or charge-coupled device.

Light-sensitive area

An Introduction to CCDs

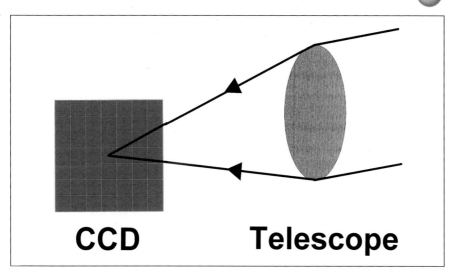

Figure 1.2 CCD at the focal point of the telescope.

Each one of these squares, which are known as pixels, is sensitive to light. Unlike film, on which an image will form that can be developed, a pixel can only count how many photons fall on it, or at least 30% of them. An alternative name for the CCD could be "clever counting device" because, reduced to basics, what it does is count photons. (In reality it counts electrons because the photons, when they react with the silicon of the CCD chip, produce electrons.) But how does counting photons (electrons) produce a picture?

Placing the CCD at the focal point of our telescope (Figure 1.2) enables it to receive the light from the object at which the telescope is pointing. In the illustration the CCD is simplified to a 7×7 pixel array, which is by our standards very crude in terms of resolution, but it will serve to demonstrate the principles.

Suppose we point our telescope with this 7×7 array at a star. Each square, or pixel, will patiently count photons of light while the exposure takes place. At the end of the exposure all we have to do is transfer those 49 counts (numbers) to a computer (Figure 1.3) where they can safely be stored. You will see in the example that where the star is located the maximum number recorded is 10 and that the count drops to 0 where there is no star – always assuming there is no light pollution, of course!

Having returned the numbers to our image-processing computer, back indoors, we are ready to display them, but because they are pure numbers

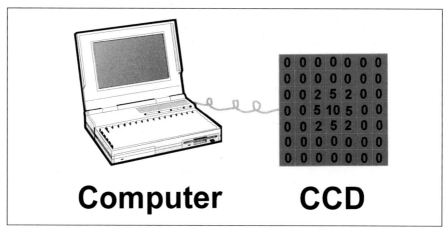

Figure 1.3
Recording photon counts.

(Figure 1.4) our brain cannot make much sense of them. However, a computer can very easily replace each number with a proportional shade of grey between, say, white for 10 and black for 0 (see Figure 1.5).

At last there is something we can recognise – a star. Well, a square one at least! Rather than proving that stars are square, this shows that 7×7 is not enough squares on our "chessboard" to delude our eyes into believing that we are looking at a real picture. If we use a higher number of squares then the eye can be fooled into perceiving a realistic image (Figure 1.6).

Figure 1.4
Displaying the counts.

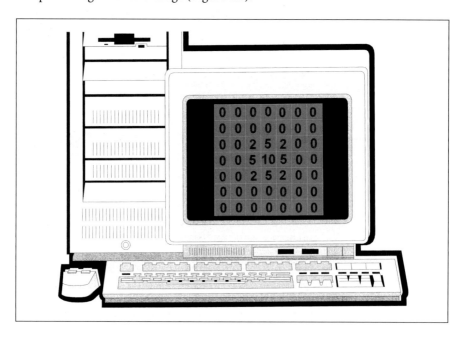

An Introduction to CCDs

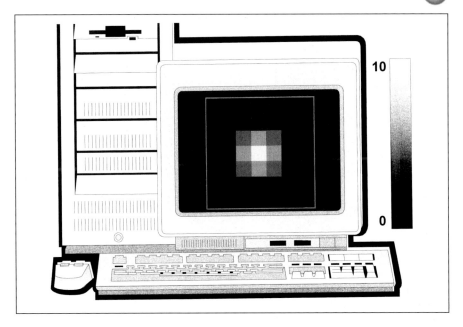

Figure 1.5
Displaying the image in shades of grey.

The stars are still square if you look with a magnifying glass, but to the naked eye the result is now photorealistic. This method of digitally recording an image is not new; space probes have been doing it for years. The Voyager spacecraft, for example, never returned any film; instead it sent back numbers to Mission Control. How many people realised that those fantastic images of the planets were actually made up of little squares? In the CCD era we are now our own Mission Control, receiving our numbers and, hopefully, with the help of this book, turning them into beautiful images.

Light Pollution

We have already dealt with some of the advantages of CCDs, namely their ten-times-better sensitivity than film, their lack of reciprocity failure, and their almost instant display on a computer screen as soon as the exposure is complete. However, the biggest problem facing the average amateur astronomer today is light pollution – the all-pervading skyglow caused by wasteful lighting in our towns and cities, in ever-increasing amounts. From many locations the interesting

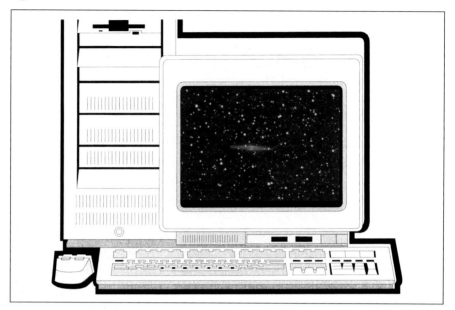

Figure 1.6 Displaying a realistic image.

deep-sky objects are now actually fainter than the night sky! CCDs cannot cure light pollution, but they allow astro-imaging from all but the worst site.

Objects fainter than the night sky can still be successfully imaged. The secret lies in the digital nature of the image. Imagine a cross-section through our CCD (Figure 1.7), again centred on an object, but this time with up to 200 levels of intensity. If we imaged from a dark sky, (a) is the kind of result we would get with an arbitrary maximum of 100 units (to keep it simple). However, in the real world we also have light pollution (b), again taken arbitrarily as 100 units. The two

Figure 1.7 The effect of light pollution on a CCD image.

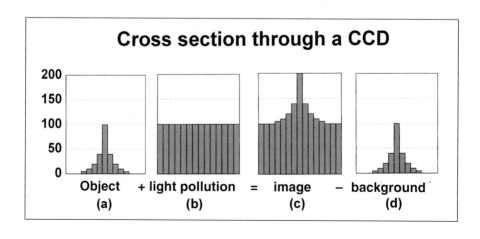

signals are additive, and the result is (c). This is why we can image objects fainter than the night sky. Wherever there is an object producing photons the result will be a bigger number than just the photons produced by the night sky. Now comes the clever bit. As our image is made up of numbers, a computer can very quickly subtract from each number any value we wish. If we subtract 100 units we are left with (d), which is remarkably similar to the dark-sky image (a).

It must be pointed out immediately that this is a gross simplification of reality. What is lost is the dynamic range available for the object. In this example, both the object and light pollution had the same maximum value, so the dynamic range available was halved. This is because CCDs have a limit to their counting ability, known as "full well capacity". If the light pollution is severe, or the object very much fainter, then the CCD will be swamped (saturated) before there is time for a meaningful signal to be received. The technical description is that the signal-to-noise ratio will be too low. So dark skies are still the best, but all is not lost under light-polluted ones.

Co-Adding Images

Assuming light pollution is present but not totally swamping the CCD, we have seen that the dynamic range available is reduced, or (more accurately) the signal-to-noise ratio will be low. The way this can be improved is by taking several shorter exposures and co-adding the images. Because our images are made up of numbers, a computer can add two images together rather easily. Although adding two images together doubles the numerical values, owing to the inverse-square law it only increases the signal-to-noise ratio by 1.4 (Figure 1.8). This is one of those unfortunate facts of physics. To double the signal-to-noise ratio we need four images.

All is not lost, however, because taking several shorter exposures rather than one long one and then co-adding images has other practical benefits. For a start, if something goes wrong during an exposure, such as an aeroplane flying through the field of view, or a neighbour switching on a (so-called) security light, or if we just bump the telescope, then only a short

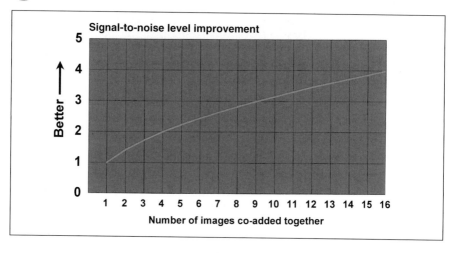

Figure 1.8 How the signal-to-noise ratio is improved by co-adding images.

exposure is lost. It is better to lose a 5 minute exposure than a 30 minute one! Other benefits are gained by randomly shifting the image on the CCD for each exposure, so that artefacts due to chip defects can be averaged out. Long exposures also increase the likelihood of bleeding, that is of "spikes" appearing on bright objects where the pixels have been saturated. Because reading out images (converting the signal and transferring the numbers to a computer) always involves a loss (noise) to the image, a camera with low readout noise is obligatory if co-adding is being contemplated.

How Many Bits?

CCD cameras are often advertised as having 8 bits or 16 bits or anything in between. What does this mean? Put simply, the electronics in the camera (the analogue-to-digital converter) will divide the full well capacity by 2 to the power of number quoted: 8, 10 or whatever. Thus the number of bits represents the maximum dynamic range possible; for example, an 8-bit CCD camera would give 256 levels (or shades of grey) to form an image (see the table opposite).

How many levels do we need? 256 seems plenty, and in fact the human eye would do well to distinguish 256 different shades. Unfortunately, the maximum range is not the same as will be available in the final image. Using an 8-bit camera might typically result in a background

An Introduction to CCDs

Number of bits	Resolution (shades of grey)
8	256
10	1 024
12	4 096
14	16 384
16	65 536

level of 150 and an image maximum of 170. We can subtract the 150, but there will only be 20 different numbers or levels left available to display, which is clearly not enough to achieve photo-realism. Thus you can now begin to see why a 16-bit camera with its maximum of 65 000 levels is not the useless overkill it might seem. However, let me give a word of caution. Any number can be divided by another number, and it must be questionable whether some 16-bit cameras are in fact producing significant numbers or just producing precise noise. My personal opinion is that 12-bit would be a minimum, with 14-bit or 16-bit a better option, if possible.

How Big? How Many Pixels?

How big should the pixels (the squares on our "chessboard") be and how many do we need? What about the physical size of the chip? All these three variables plus, unfortunately, cost, are interrelated. There is no more emotive topic among CCD users than this! There has been endless debate in the Astronomy Forums on the Internet concerning pixel size and matching it to resolution and image size. Suffice it to say opinion is divided.

To take a simpler approach, let us start with how many pixels are required to produce an image that passes as photo-realistic. Where are the created images going to be viewed? The answer is: probably on a computer screen. So the question could be rephrased as: at what point does a computer screen become photo-realistic? The original VGA standard of 640 × 480 pixels was not, hence the emergence of the Super VGA (SVGA) standards. I think most would agree that 800 × 600 is the minimum to achieve near photo-realism on a screen, with 1024 × 768 ideal. Therefore, if "pretty pictures" are the goal (and I see nothing wrong in that – it is a hobby after all) then a CCD camera with around 800 × 600 pixels would be a good choice. If one is supernova hunting or if the planets are the target then a chip with fewer pixels (say 300 × 200) could well be sufficient. Bear in mind, however, that a bigger CCD chip does make finding and centring easier.

Now to the delicate subject. How big should the pixels be? An analogy can be made with film, where fine grain equals high resolution but low speed, whereas large grain equals low(er) resolution but high speed. It may be worth pointing out at this stage that the fine-grain film widely used in astrophotography, Technical Pan 2415, has slightly higher resolution than the 9 μm pixels of the Kodak chips. However, that "fine grain" comes at a price of lower speed, so it would be wasteful to use a resolution which the telescope itself is incapable of delivering.

For deep-space images, a rule of thumb is that for focal lengths up to 1500 mm (perhaps 2000 mm with excellent optics and seeing) 9 or 10 μm pixels are sensible. Beyond this focal length, larger pixels come into their own. Bear in mind that some cameras can be operated in "binning" mode, that is, pixels can be electronically combined to produce bigger ones. With 2 × 2 binning the Kodak KAF1600 chip reduces to 768 × 512

Figure 1.9 Image appearance comparing 9, 18 and 27 μm pixels for a focal length of 1250 mm.

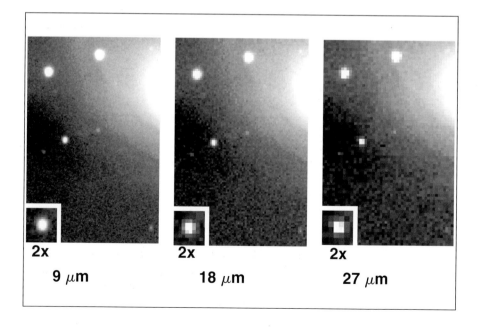

An Introduction to CCDs

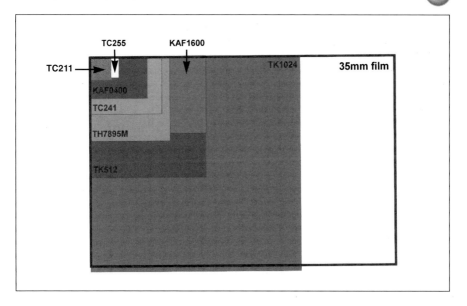

Figure 1.10
CCD chip sizes compared with 35 mm film.

pixels with an effective size of 18 μm each, making it a better match to focal lengths in the range 2000 mm to 3000 mm. The best thing to do is to look at the images in this book and decide for yourself. Figure 1.9 shows part of a galaxy (M81) and the effect on image quality for three different pixel sizes.

How big physically does the CCD need to be? The easiest way to illustrate this is to compare the various sizes with 35 mm film, to which most people can relate (Figure 1.10). However, decisions on pixel size and number will already have limited the choice. What really decides physical size is the type of objects to be imaged, and of course the cost, always assuming you have already got a telescope. For example the TC255 and the KAF0400 chips have similar pixel sizes, but to use the former for deep-space objects would require a much shorter focal length than it would to use the latter, simply to fit them on the chip. All is not lost with small chips, as a mosaic can be assembled from several images to cover a larger object. At the back of this book, Appendix C gives the field of view for common CCD chips for a variety of focal lengths, and Appendix D the sizes of some common deep-sky objects. One final point: the higher the total number of pixels, the bigger the files and the slower the processing. There is no point in having that megapixel CCD if a single image fills your computer and your monitor is not capable of showing it!

Calibration

Before image processing can be undertaken, the image must first be calibrated. That is, we need to correct for defects in the actual image taking process. Image processing will bring out and emphasise the fine, almost hidden detail in the image. We need to be sure that the details being revealed are real, and not defects originating in the optics or camera.

So far we have concentrated on the advantages of CCDs, but they have defects too, namely "dark current" and "non-uniformity". However, with careful calibration both can largely be eliminated. Let us take dark current first. This is the CCD's sensitivity to heat and manifests itself by recording a signal, even if the telescope is capped! It is in fact recording a heat signal which, like light pollution, eats into the dynamic range of the CCD. That is why CCDs are (usually) cooled, and those with lower dark current are best for our use, where long exposures are the norm. If the CCD has regulated cooling then the simplest way to correct for dark current is to take twin equal-length exposures – one of the object and another with the telescope capped. The two are then subtracted leaving only the signal from the sky. If the CCD does not have regulated cooling, then a before and after dark frame, as they are called, can be averaged and then subtracted. With more sophisticated cameras there are better ways of eliminating dark current, usually involving a master dark frame and a separate bias frame (dark frame of zero exposure length). These are techniques that can be mastered later, and a simple dark-frame subtraction is capable of giving excellent results, albeit somewhat wasteful of observing time.

Non-uniformity (the inability to record a uniform object uniformly) can also be corrected. It results from two sources. One originates in the telescope and the other in the chip itself. The common non-uniformity from the telescope is "vignetting", where focal-plane illumination is not uniform and reduces towards the edge of the field of view. Schmidt–Cassegrain telescopes, particularly those with focal reducers, are prone to it (see Figure 1.11). The non-uniformity from the CCD is due to different sensitivities of each individual pixel and other problems, such as dust on the optical window. The good news is that both can be corrected by means of "flat fields". The tendency is for the

An Introduction to CCDs

Figure 1.11 Flat field produced by Schmidt–Cassegrain telescope with focal reducer.

beginner to ignore flat fields, but often they make a critical difference to subsequent image processing.

Not surprisingly, flat fields are produced by pointing the telescope at a flat (uniform) field such as a piece of white paper, a dusk sky or clouds at night if you have light pollution – so the last mentioned does have its uses after all! If the telescope plus CCD was perfect, the result would be a totally uniform image but, as such a combination has never been made, it won't be. The flat field should in fact be the median of several flat fields (the more the better), each with dark frames subtracted. The flats should be "normalised", that is, brought to the same average brightness, before the median is taken. Images are then divided by the flat field. They need to be taken for the exact orientation of the CCD during the night's exposures and ideally at the same focusing position too. If filters are used for tricolour (i.e. red–green–blue), or for blocking light pollution, then these should be in position when the flats are taken. If you are co-adding images, then each individual image needs to be flat-fielded before adding. Once dark-subtracted and flat-fielded, the image is calibrated (see Figure 1.12) and the exciting image processing can begin.

Image Processing

The subject of image processing is too vast to cover fully and is of course discussed by each contributor in turn, so only an introduction to the essentials is given. Digital image processing is the equivalent of darkroom

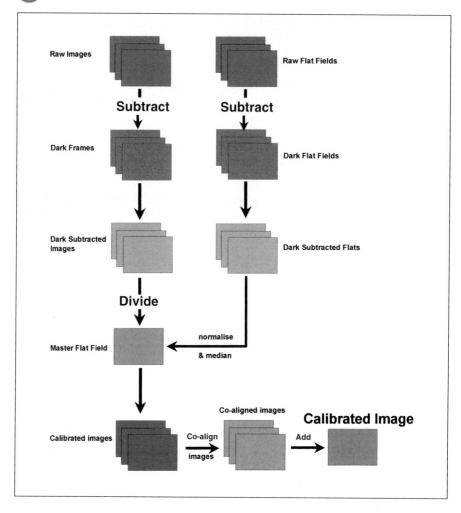

Figure 1.12
Image calibration procedure.

magic in the photography days. Instead of chemical and exposure tricks, all we need is a computer and some clicks with a mouse. However, first there are some truths to be stated. Image processing cannot turn a poor image into a good one. It merely makes the best ones better. The better the focusing and the better the guiding and the better the signal-to-noise ratio then the more an image can be improved.

Sensible image processing will involve subtracting sky background, filtering to reduce noise and usually "log" scaling (i.e. the logarithm of each number in the image is taken). This will produce a black sky and lift faint details to make them visible. For planetary

An Introduction to CCDs

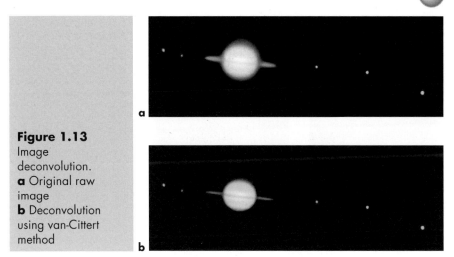

Figure 1.13
Image deconvolution.
a Original raw image
b Deconvolution using van-Cittert method

images, old favourites such as unsharp masking, the digital equivalent of David Malin's darkroom wizardry, take some beating. Master the basics before attempting the exotic.

The restoration of images to correct for the inevitable defects (such as atmospheric turbulence, guiding inaccuracies and telescope deficiencies) is known as deconvolution (see Figure 1.13). It probably came to most people's attention when the first Hubble Space Telescope images proved to be fuzzy because of spherical aberration. Deconvolution of the digital images by some of the most powerful computers on Earth was able to rescue those early images. However, such image processing needs to be carefully carried out, particularly with amateur images where the signal-to-noise ratio is often not as high as it should be. There is a lot of hype over exotic processes, such as "maximum entropy deconvolution", the results of which can be spectacular but which can also have side-effects. You have probably seen images with black haloes around bright stars (Figure 1.14) – these are artefacts of image processing and indicate too enthusiastic a processing, to put it politely!

Study the variety of image processing techniques used by the contributors to this book. There are many excellent images shown. Forget the hype and make your own mind up as to which will be most applicable to your own circumstances.

 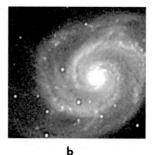

Figure 1.14
Excessive image processing – dark haloes. Comparison of log scaling (**a**) and excessive deconvolution (**b**). Note that the stars are smaller in (**b**) but they have dark halos where they overlay the galaxy.

Colour Imaging

CCD cameras are monochrome (with the single exception of a Starlight Xpress model). That is, although they are very good at counting photons, they do not differentiate between colours. All is not lost, however. Probably the best colour astrophotographs in the world, those of David Malin, are in fact taken with black-and-white film. His technique of shooting three exposures through red, green and blue filters, and then recombining them in the darkroom to produce colour prints, is equally applicable to CCD colour imaging. Not only that; we do not need David's darkroom skills, because a few clicks with a mouse will do the same for us. The major problem to be overcome in tri-colour imaging is the poor blue response of many CCD chips, particularly the first series Kodak KAF range (the E series is claimed to be up to four times better at 400 nm). This can result in very long blue exposures. Another point to be aware of when using filters is to ensure that they are all of the same thickness, so that they will all focus at exactly the same point. Refocusing between exposures is not practicable and should be avoided, particularly for an object spinning fast such as Jupiter, where there simply will not be enough time.

One complication for us is that, as well as being sensitive to visible light, CCDs are remarkably sensitive to infrared (IR). There are differences of opinion as to whether an inline IR (blocking) filter should be used if taking tri-colour images through filters. In theory it should, but whether it makes a noticeable difference depends on what wavelengths the object emits. IR sensitivity opens up new opportunities, however. It is popular with professionals, some of whose work is tri-colour in IR, red and blue-green. With CCDs this is no problem. Equally, some objects emit strongly in the IR, and so these are now in our range. IR-passing filters, which look black to the eye, are available from most good camera dealers.

There is another form of colour imaging, referred to as "false-colour", by which grey-scale images can be given subtle shades or exotic colours to highlight details. In the Colour Gallery there are examples of both, and if the former is carefully done then the result is a remarkably realistic image.

The Future

Although we have concentrated on being able to image fainter objects and/or in greater detail, that is not the only use for which CCDs excel. Photometry and astrometry are areas where CCDs can do real work. Photometry is the measurement of the brightness of stars. The linear response of CCDs, their dynamic range and stability of the sensitive area (compared with film) are characteristics that make them so powerful. With good calibration and signal-to-noise ratios, magnitudes can be calculated to a hundredth of a magnitude, or better! Variable-star enthusiasts will at once realise the powerful tool that this represents.

Astrometry is the measurement of position, and was traditionally carried out using precise mechanical measuring engines, a laborious process at best. With a CCD and a good reference source of stellar positions, such as the *Hubble Guide Star Catalogue*, the position of a newly discovered object can easily be calculated, along with its orbital elements. These are compared with those of known stars in the image; three stars are the minimum, but the more that are used, and the greater is their spread around the object, the better the results will be. This is now the standard method used for new comets, and when they are first spotted their orbital elements are invariably calculated from CCD images.

For amateur astronomers, the CCD offers a new dawn with a tool so powerful that only a few years ago a telescope measured in metres would have been needed to emulate the results that are now attainable from backgardens. The next chapters give an indication of what has become routinely possible thanks to the silicon marvel that is the CCD.

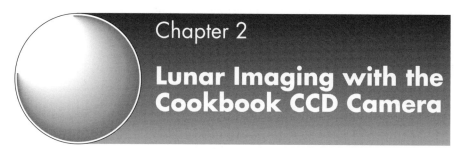

Chapter 2
Lunar Imaging with the Cookbook CCD Camera

Dave Petherick

Introduction

The CCD revolution in amateur astronomy has been spurred on in part by the release in 1994 of *The CCD Camera Cookbook* by Richard Berry, John Munger and Veikko Kanto (published by Willmann-Bell). For those with the desire and some soldering skills a CCD camera can be built at modest cost, which for some may represent the only hope of entering this arena of amateur astronomy.

For me, the interest was not only in getting a high-quality CCD camera for the lowest possible price, but also in learning about its operation and understanding its capabilities and limitations. Using the circuit boards purchased along with the book, and utilising a friend's machining expertise for the camera housing, I completed my "Cookbook CCD" in August 1994. It is based on the Texas Instruments TC245 CCD frame transfer chip, which eliminates the need for a shutter on bright objects. The contents of the image area are quickly shifted to a light-protected storage area. This prevents blurring and/or streaking of the bright object(s) during the readout phase. Exposures ranging from 0.001 to 999.9 seconds are possible. In the spring of 1995, Willmann-Bell released improved camera control software (245PLUS) and instructions for modifying the camera to operate in "low dark current" (LDC) mode. I completed these modifications in the summer of 1995.

There are several camera operating modes that make the camera easy to use. The basic CCD chip size is 6.4 mm × 4.8 mm and 378 × 242 pixel mode (2 adjacent pixels binned together) or 252 × 242 pixel mode (three adjacent pixels binned together). In this way, camera sensitivity is improved at the cost of some resolution. This brings up the important subject of matching the telescope focal ratio to the mode in use to maximise the resolving power of the telescope – CCD camera combination. The Nyquist sampling theorem requires two pixels for each unit of resolving power of the telescope. For example, if the telescope can resolve 1 arc second, we choose a telescope configuration that ensures two pixels per arc second if our goal is to maximise the image resolution.

For lunar and planetary imaging, where the subject is relatively bright, this criteria is met at a focal ratio of about $f/40$. On my home-built 8 inch (20 cm) $f/8.2$ Newtonian reflector telescope, I use a 12.5 mm ($\frac{1}{2}$ inch) orthoscopic eyepiece with an eyepiece projection adapter to achieve this, and even higher focal ratios. This is the basic approach for all planetary CCD imaging. For wide area lunar images, prime focus can be used. The problem is that the moon is extremely bright!

The solution is to use dense eyepiece filters on the CCD camera to reduce the intensity of the light. CCD cameras are very sensitive, and if too much light hits the CCD chip, then a process called "blooming" will result. Pixels that have been completely filled with electrical charges (resulting from photons of light striking the chip) overflow their charges into the surrounding pixels, leaving trails in the image in the direction in which the chip readout occurs. If the entire chip is "saturated", all you have is a pure white image! Learning to use a CCD camera invariably calls for experimentation. You try different configurations, filters, and exposure times until you strike a balance between subject brightness, atmospheric seeing conditions, and image quality. With experience, you learn the appropriate setup for any given subject, observing goal, and sky condition.

Setup

My most convenient observing site is my own backyard. As this is in a rural subdivision, the skies are pretty good except for the local streetlights, which illuminate the site

Lunar Imaging with the Cookbook CCD Camera

as if it were daytime! This may be hard on the night vision, but CCDs seem to be able to overcome such adversity as long as the open end of the telescope is well shielded from stray light. My equipment consists of a home-built 8 inch (20 cm) Newtonian reflector with 1.3 inch (33 mm) diagonal mirror on a Cave Astrola equatorial mount (see Figure 2.1). A Celestron single-axis drive corrector is used to stabilise the polar axis drive, allowing fine telescope aim adjustments without touching the telescope. A JMI Motofocus electric focus motor allows fine tuning of the focus without shaking the telescope. A manual declination adjustment is also provided. A Telrad unity power finder is used for rough alignment and a Celestron 60 mm (2.4 inch) Gregorian telescope acts as a medium 33-power finder to allow quick placement of a subject in the CCD field of view.

Figure 2.1 Dave Petherick and his home-built 8 inch (20 cm) Newtonian reflector on a Cave Astrola equatorial mount. His light box is on the ground, front right.

The Cookbook CB245 camera requires some support equipment: a power supply for the camera electronics, a thermoelectric cooler chip, and a coolant circulation pump. The closed-circuit cooling system consists of an electric bilge pump to flow the coolant (a mixture of water and isopropyl alcohol) through the heat exchanger on the back of the CCD camera. This, along with the Melcor thermoelectric cooler, transports heat from the CCD chip (which allows it to cool to about 30 °C below ambient) and rejects it to a water bath through a coil of copper tubing.

The camera head (See Figure 2.2) includes a high-quality, coated optical window in front of the CCD chip and a standard 1.25 inch (32 mm) extension to fit a standard telescope eyepiece adapter–focuser. It incorporates preamplifier electronics that connect to an

Figure 2.2 The camera head of the Cookbook CCD camera built by Dave Petherick.

interface box about 8 inches (20 cm) away, which is in turn connected to the parallel port of a PC-compatible computer via a 15 ft (4.5 m) ribbon cable. The 245PLUS software controls the camera operation by reading and writing to the parallel (printer) port.

Various optical accessories are used to allow Barlow projection, eyepiece projection and use of colour, and also neutral-density and infrared-blocking filters (alone, or in combination), to achieve a number of optical configurations.

Taking Images

Once setup is complete, imaging can begin! Prime focus images of the moon require several filters to allow exposures of about 0.3 second. This usually results in the full use of the camera dynamic range which is 4095 grey levels (12 bit) for the Cookbook 245 (CB245) camera. Since the images are in digital format, large mosaics of the Moon or even of deep-sky objects can be assembled using appropriate software.

My favourite image is the full-Moon mosaic (Figure 2.3), which was assembled from 15 of 38 original 252 × 242 pixel prime focus CCD images. A stack of every filter I owned at the time (September 1994) (0.6 ND, 0.9 ND, #47 violet, #24A red, #80A light blue) was used to attenuate the lunar brightness. This filter combination, combined with the infrared sensitivity of the CCD chip, means the image is primarily taken in the near-infrared spectral region. Each of the images employed a 0.3 second exposure time.

The CCD chip was aligned with the telescope axes to make the process of scanning across the Moon with sufficient overlap of the images as easy as possible. I scanned across the Moon starting at the northern hemisphere, then shifted down and back across until the South Pole was imaged. I repeated the process back from south to north just to make sure I had not missed any of the lunar surface.

Image Processing

The individual images were calibrated (dark-frame subtraction only) and then processed using unsharp

Figure 2.3
Full-Moon mosaic assembled from 15 of 38 original 252 × 242 pixel prime-focus images. Taken with Cookbook CCD camera and 8 inch (20 cm) Newtonian.

masking in Richard Berry's CB245PIX (now called CB245). A triangular unsharp mask was applied in two steps to get the best resolution without introducing excessive high-frequency noise in the images.

The most important requirement was not to do any brightness scaling at this point. When the mosaic segments are aligned, slight tonal corrections are necessary to provide a perfect, seamless match. Any previous scaling operations will make it difficult, if not impossible, to match the dark and light tones at the four boundaries of each frame. Because my flat-fielding technique was not well developed at that time, it was necessary to trim each frame using the CB245 CROP function to eliminate any edge anomalies after processing. (The TC245 CCD chip has a mask over the light-sensitive surface that can partially block the pixels at the extreme edges, leading to an edge column or row of bright pixels; flat fielding normally compensates for this effect.)

After aspect ratio correction (to yield square pixels) and conversion to TIFF format, the mosaic was assembled on the computer using Aldus PhotoStyler 2.0 for Windows. The best fifteen frames were carefully pasted, aligned and overlapped, and slight tonal corrections made, to ensure a perfect, seamless match. The resulting image was about 800×800 pixels.

Lunar Close-ups

High-resolution images using eyepiece projection are possible when the atmospheric seeing is good. I have tried short-duration exposures (as low as 0.001 second) to "beat the seeing", but the images are never as noise-free as longer exposures (0.3 to 1.0 second) under a calm sky that employ the full dynamic range of the camera. Such images have a wide range (2000 to 3000 pixel values) and respond well to image processing techniques such as unsharp masking and linear brightness scaling.

Images of Plato (Figure 2.4), Clavius (Figure 2.5), the Lunar Domes (Figure 2.6) and Copernicus (Figure 2.7) all show the potential of the CCD in making near-observatory-quality lunar images from your own backyard.

Image Calibration Tips

Dark Frames

The typical image calibration process includes dark-frame subtraction and flat fielding to yield a linear representation of the light hitting the CCD chip. The dark frame is basically an exposure of the same duration as the image frame, except that no light is allowed to enter the camera. The easiest way to achieve this is to cap off the end of the telescope. This eliminates the possibility of changing the camera orientation prior to taking the flat-field frame. The dark frame consists of a combination of bias (offset from zero pixel value) plus any

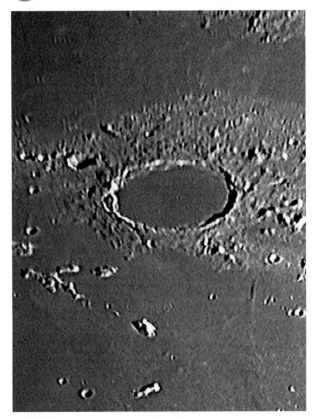

Figure 2.4
Plato.

thermal noise. Thermal noise is present in any CCD exposure. This noise is minimised in the CB245 camera by thermoelectric cooling of the CCD chip using a Peltier ceramic cooler device. The low dark current (LDC) modification also helps minimise thermal noise for most pixels, but should be turned off for flat fields, because some pixels remain "hot" and will add noise to the final image.

Flat Frames

The flat field is an image of a uniformly illuminated surface taken with all optical elements in place (i.e. same camera orientation, f-number and focus) with an exposure duration which yields a block of midrange pixel values (2000 to 3000 for the CB245). I constructed a light box (Figure 2.8) for making flat fields and use CB245 to do the dark subtraction and averaging functions.

Lunar Imaging with the Cookbook CCD Camera

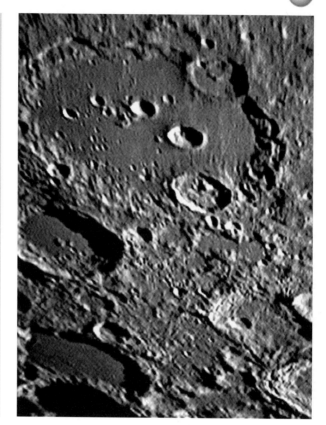

Figure 2.5
Clavius.

The box contains four low-wattage incandescent bulbs connected to a dimmer. The box is basically an integrating cube whereby the light is directed onto the inside surface (painted flat white) and scatters as uniformly as possible. The light then passes through a sheet of milk plastic (as used in fluorescent light diffusers). The light box attaches to the tube stiffening ring at the open end of my 8 inch (20 cm) Newtonian telescope. The light box can be rotated to yield any number of flat fields which, when averaged, results in an excellent master flat.

Equivalent dark frames are also taken at the same exposure length, averaged and subtracted from each flat field. All of the resulting flat fields are then averaged to yield a final master flat frame for the given optical configuration. By dividing the dark-frame-subtracted image by the flat field, any slight variations in individual pixel sensitivity are compensated for. Noise in the final image is reduced by averaging a number of dark frames and flat field frames, giving a

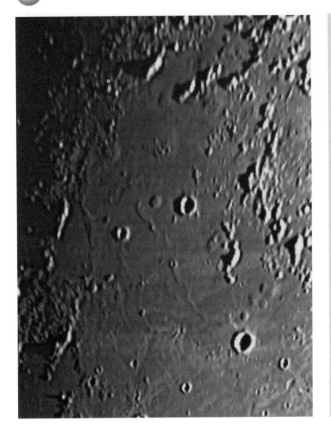

Figure 2.6 Lunar Domes.

completely calibrated image. The image is now ready for image processing to extract the maximum amount of detail present.

A Typical Imaging Session

In a perfect world, every time you wanted to observe and make CCD images, all the equipment would be set up and operating the moment you arrive at the observatory. Well, in the real world, I haul my rather heavy Cave Astrola equatorial mount, 8 inch (20 cm) $f/8.2$ home-built Newtonian reflector, Cookbook CCD camera and cooling system, power supply, computer, monitor, optical accessories, star charts, etc., and set up in my backyard! Now that this is routine, I'm usually ready to image in about thirty minutes.

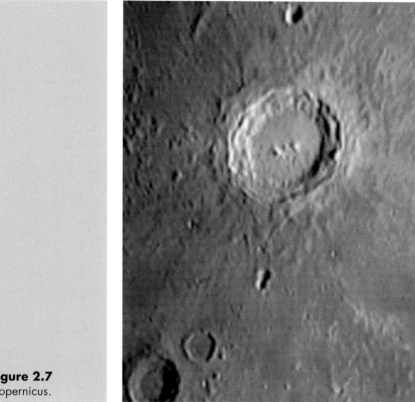

Figure 2.7
Copernicus.

I always find it worth the effort to do an accurate polar alignment in case I do some deep-sky exposures. I can fine tune the alignment by checking for declination drift watching the computer monitor as the camera images a star field in "find object" mode with a one- or two-second exposure. The 245PLUS software allows you to select this "find object" mode in which 12 adjacent pixels are binned together to give a small but extremely sensitive, full-frame-binned to one-quarter size, 126 × 121 pixel image. Bright deep-sky objects are easily visible on the computer screen in near real time (exposures of 2 to 5 seconds). A "focus" mode quickly reads the centre 126 × 121 pixel area of the chip. I find the best focus method is to image a moderately bright star at an exposure that results in minor blooming. The focus is fine-tuned until blooming length is maximised (I use an electric focus motor so the telescope remains stable during focusing).

The ability to balance the telescope is also important as one changes from prime-focus imaging to eyepiece

Figure 2.8
Home-made light box for taking flat-field images fitted to the telescope front.

projection to searching for an elusive deep sky object using an eyepiece which is parfocal with the CCD camera. Reducing any mount or telescope flexure is also important, especially when doing eyepiece projection where the field of view is very small.

I aim the telescope at a bright object at moderate power, centre the object and engage the drive. The Telrad and medium-power finder are then aligned with the telescope. I use the Telrad unity-power finder to get close to my subject, then switch to the medium-power (33×) finder (a Celestron C60 Gregorian spotting scope with a crosshair eyepiece). By now the object is usually in the CCD field of view and can be positioned or centred on the chip using the slow-motion and drive corrector controls.

By observing any drift in the image on the computer screen, you can detect any imbalance, polar alignment, seeing effects or other problems that will affect the observing session. Items under your control must be corrected now or you will simply take a number of fuzzy, blurry images.

The 245PLUS software allows you to take multiple exposures automatically. This is very convenient for taking a number of short-exposure images, dark frames and flat fields. In this way, you can catch the moments of excellent seeing and select the best images for further processing and discard the bad ones.

I take a number of dark frames throughout an observing session, because the camera is sensitive to temperature changes resulting in bias drift during the night. The latest versions of CB245 now include dark-frame matching. This is a technique whereby you make a dark-frame of the longest exposure you intend to make that evening, and the software will automatically scale the dark frame for all shorter-exposure images during the image calibration process. The 245PLUS software also includes a "drift subtract" option to compensate for bias drift. I will be using this technique from now on.

It is also a good idea to check focus a few times during the night, to ensure that you are taking the best possible images. Soon it is time to call it a night, disconnect the equipment, haul it back inside and head for bed. The next day you can review your images and start calibrating and image processing to see what treasures have been captured! You can try a number of techniques to see which one gives the best results. Don't be afraid to experiment!

Conclusion

The CCD revolution gives you the power to image the Moon, planets and deep-sky objects as never before, and you can spend those cloudy nights refining your image processing skills! Don't forget to look through the eyepiece every once in a while! The joy of visual night sky observing need not be sacrificed – it is only enhanced by these new abilities to enjoy the universe in which we live!

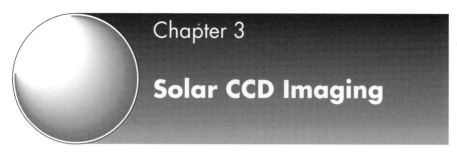

Chapter 3

Solar CCD Imaging

Brian Colville

Introduction

I have been actively involved in amateur astronomy since I was a young boy. After twenty years of observing my interests have centred on observing, photographing and imaging members of our solar system. I first entered the CCD field in 1991 with the purchase of a Lynxx PC Plus CCD camera, and quickly discovered that this was ideally suited to solar, lunar and planetary imaging. Although the Sun was the last target of my CCD observations, it certainly has been one of the most satisfying objects of study.

Setup

Location

My observing site is located on our family dairy farm, approximately 100 km north-east of Toronto, Ontario (44° 17′ N, 79° 10′ W) (Figure 3.1). This site has the advantage of being far removed from large urban heat sinks, and it is 6 km inland of Lake Simcoe.

Prevailing westerly winds and an onshore breeze from the lake mean moderate air temperatures and steady the seeing during the summer months. The best

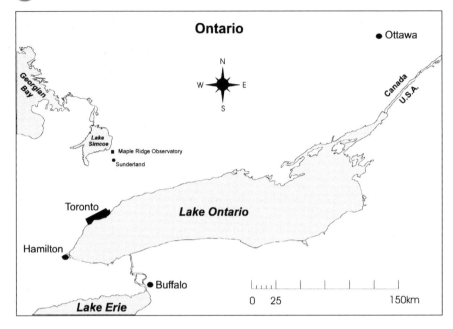

Figure 3.1 My observing site.

conditions for solar imaging usually exist during the mid-morning when the Sun has climbed high enough into the sky to escape the impact of atmospheric effects near the horizon, and solar heating has not introduced instability to the airmass above the telescope. Most of my solar work takes place on weekends when I can spend time at the telescope in the morning. During the summer I sometimes get steady seeing during the late afternoon, when my view of the Sun extends to the north-west over the lake and temperatures start to fall; it is not as good as observing in the morning, but offers more chances to observe on days when I have to be at work.

Equipment

CCD Camera

The CCD camera which I use for solar imaging is a Lynxx PC Plus, manufactured by Spectra Source Instruments. This utilises the TC211 chip. It has medium-sized pixels, 13.75 × 16 μm, arranged in an array 192 pixels × 165 pixels, forming a chip which is 2.64 mm square. The Lynxx camera features an unregulated two-stage thermoelectric cooler which maintains the CCD temperature approximately 40°C below the ambient air temperature. The image is digitised to 12 bits per pixel, yielding 4096 shades of grey.

The Lynxx camera consists of two parts. The first is the camera head, which contains the CCD chip, mechanical shutter, and thermoelectric cooler. The shutter is capable of exposures as short as $\frac{1}{100}$ second. The second component is a circuit board, which plugs into an ISA expansion slot on an IBM-compatible computer. This board controls the camera head, digitises the image, and interfaces with the control software. The card also produces a NTSC/PAL video output, which can be displayed on a composite monitor. The video output is refreshed at a faster rate than the VGA image on the host computer: up to 3 frames per second in subframe mode. This allows focusing to be done quite easily compared with some cameras, which can take several seconds to download an image through a serial cable. The control software provides for focusing, image acquisition, basic image processing, image calibration, and saving images to disk.

Telescope Hardware

The primary telescope which I use to image the Sun is a 25 cm (10 inch) Meade Schmidt–Cassegrain reflector, and a 10 cm (4 inch) Celestron refractor is used as a photo-guidescope. Full-aperture Thousand Oaks Type II solar filters are used on both scopes. The refractor is used to centre the main scope on the targeted sunspot group and monitor the seeing for the best imaging opportunities. I occasionally use the refractor for imaging when the sunspot groups are too large to fit in a single frame when imaging with the 25 cm telescope. Both telescopes are carried on a Losmandy GM100 equatorial mount (Figure 3.2).

The Thousand Oaks filters reduce the brightness of the Sun sufficiently for visual observation and conventional photography, but the image is still too bright to image with the Lynxx camera. In order to reduce the brightness to a level which does not saturate the CCD chip, I use Wratten 25 (red) and ND 6 (neutral-density) filters between the telescope and camera. The coloured filters are carried in a manual filter wheel, and the ND

Figure 3.2 Ready to go: 25 cm and 10 cm (10 inch and 4 inch) telescopes with solar filters on a Losmandy GM100 mount.

filter is threaded into the nosepiece of the camera (Figure 3.3). No infra-red blocking filters are used.

This filter combination allows me to record the Sun without saturating the image with exposures between 0.01 and 0.05 second. Short exposures are essential for solar imaging if you hope to "freeze" the effects of atmospheric turbulence, and capture the maximum detail that your telescope can deliver. CCD cameras which have mechanical or "frame transfer" shutters

Figure 3.3 Lynxx camera head with neutral-density filter and filter wheel assembly.

capable of exposures of $\frac{1}{100}$ second or less would be the best choice for solar imaging.

The Lynxx camera field of view is 3.6 arc minutes square with the 25 cm telescope at $f/10$, and 9 arc minutes square with the 10 cm telescope at $f/10$, yielding a resolution of 1.25 arc seconds per pixel and 3 arc seconds per pixel, respectively. These are small fields of view by conventional film standards, but they nicely frame many of the large, complex sunspot groups. Images captured over several days will demonstrate the dynamic changes which take place on the surface of our star.

Computers

I use two IBM-compatible PCs for image acquisition and processing. The image acquisition computer has a 286 12 MHz CPU with 1 Mb of RAM, a 40 MB hard disk, a VGA graphics card and a VGA monitor. This system runs under DOS and controls my Lynxx and ST4 cameras. This computer is housed in a large, insulated tool cabinet to protect the CPU from sudden temperature fluctuations (Figure 3.4). I leave the CPU running twenty-four hours a day during cold weather, generating enough heat to keep the insulated enclosure above the dew point. The VGA monitor and composite focusing monitor ride on top of the cabinet. Casters on the base allow me to move the entire computer centre to any convenient viewing position in the observatory. The cabinet also holds most of my telescope accessories and ST4 autoguider. At the end of an imaging session I copy the images to disk and take them home for processing.

I perform my image processing on a 486 DX33 PC. This system has 8 Mb of RAM, an 800 Mb hard disk, an SVGA graphics card and monitor, and operates under Windows. The software packages I use most often are BatchPIX and Aldus PhotoStyler. BatchPIX is used to calibrate images and perform unsharp masking to enhance details visible in the images. PhotoStyler is used to give a final tweak to the images' brightness and contrast or to composite multi-frame images. All original calibrated images and the best processed images are archived to disk and backed up onto tape for future reference.

A final piece of hardware which I use is an Epson Stylus Colour II inkjet printer. It is capable of

Figure 3.4
Imaging centre with computer, monitor (*left*), and focusing monitor (*right*).

producing 720 DPI monochrome and colour prints, providing an excellent hardcopy of a day's (or night's) work.

Observatory

All of my astronomical equipment is housed in Maple Ridge Observatory, my roll-off roof observatory (Figure 3.5). This facility allows me to maximise the time spent at the telescope. My wife, Sandy, and I live in an apartment in the town of Sunderland, approximately 10 km from the farm. The observatory is just a short drive away. The permanently mounted telescope and computer require no setup time and all of the optical components are close to the ambient air temperature. Very little time is needed for the thermal mass of the tele-

Solar CCD Imaging

Figure 3.5
Maple Ridge Observatory.

scope to reach equilibrium with the air. Cameras, eyepieces and filters are all within close reach while I am working at the telescope.

Image Acquisition

Each imaging session begins in much the same manner. Once inside the observatory, I turn on the computer (if it is not already on), start the Lynxx control software, and turn on the cameras' cooler. While waiting 15 minutes for the CCD cooler to stabilise, I can open the observatory roof, install the solar filters and filter wheel on the telescope. A quick glance through the eyepiece will reveal what features the Sun has to offer that day. A final step is to check that the guidescope is aligned with the main telescope.

The first step of imaging is to focus the CCD camera. I have found that my 35 mm camera and CCD camera have close to the same focus. First the telescope is focused on the Sun through the 35 mm camera, then I switch the Lynxx camera with the regular camera body. The Lynxx camera accepts a standard T-ring and prime-focus camera adapter, which is inserted into the eyepiece holder of the filter wheel. Next, I centre the limb of the Sun in the guidescope, and rotate the red filter into position in the filter wheel.

The Lynxx camera has two focus settings. One is a "full-frame" mode and the other is a "subframe" mode. The entire image is displayed in the full-frame mode, and the centre quarter frame is displayed in the sub-frame mode. Activating the full-frame mode displays the limb of the Sun on the focusing monitor, and focus is adjusted until the limb appears as a sharply defined line. Final focus is checked by centring on a sunspot group, and making fine adjustments while viewing in the sub-frame mode. In this mode, images are updated two to three times per second. I have to watch for the sharpest image, as they tend to wash in and out of focus as the seeing changes. Once I am satisfied that the best focus has been achieved, I move the telescope to the first sunspot group that I wish to image.

To begin the process of capturing images, I take test exposures to find the exposure which saturates approximately three-quarters of the CCD's dynamic range. This will give the brightest portions of the image a value of 3000 to 3500 out of the 4096 possible grey levels. A function in the analysis menu allows the user to measure the value of a group of pixels, and the value is displayed on the computer monitor. Once the exposure time is set, I can begin recording images. The Lynxx camera has six image buffers, but only one buffer can be displayed at a given time. When an image is exposed and digitised, it is stored in the current image buffer, and new images overwrite the previous image until I select another buffer. Each time I press "return" on the keyboard, a new exposure is taken and displayed. I keep taking exposures until a crisp image appears, then switch to another buffer and repeat this process until all six buffers are filled. As few as 10 or as many as 50 images may be exposed in order to capture 5 or 6 good images. These images are saved to disk for processing later in the day. Then I move the telescope to the next target, and repeat the imaging routine until all of the features I wish to image have been saved to disk. Dark and bias images are captured throughout the imaging session, and the exposure time is also checked to ensure that the image's dynamic range is still in the target range. The final step in image acquisition is to capture the final set of calibration images, the flat field. I use the daytime sky to record the flat field images by pointing the telescope away from the Sun and removing the solar filter. Six to ten flat-field images are captured before the telescopes and other equipment are put away, and the observatory is closed up. Before leaving the observatory, I copy all of the images to floppy disks and take them home for processing.

Image Processing

Image processing is a two-stage process. The first stage is to calibrate the raw solar images, removing bias and thermal noise and to correct optical defects (vignetting, dust, uneven illumination, etc.) by making a flat-field correction. BatchPIX is used to average each group of calibration images made earlier at the telescope, creating intermediate bias, dark and flat-field images. The bias image requires no further processing. The bias frame is subtracted from the intermediate dark frame to produce a master thermal frame. The intermediate flat frame has both the bias and thermal frame subtracted from it to produce a master flat-field image. The master flat, thermal and bias frames are used to calibrate the raw solar images.

Each of the solar images are read by BatchPIX, the bias, thermal and flat-field corrections are made, and the calibrated images are saved to disk. This preserves the full dynamic range which was captured in the original image and the data are available for any new or improved processing routines. (You can always reproduce that terrific image from the raw data, but if you only keep the processed images it is impossible to go back to the original and process it another way.) These calibrated images are used for image processing and archived for future use.

I have tried many processing routines since I began imaging, but I have settled on a number of unsharp masking routines found in Richard Berry's BatchPIX software. BatchPIX is an automated image processing package which applies an image processing script to a list of images which I specify. I can use scripts which are supplied with the program, or create my own custom routines. For solar images, I use two unsharp masking routines to enhance the details captured in the images. A binomial unsharp mask with a radius of six pixels is first passed over the image, enhancing medium-scale detail in the image. A regular unsharp

Figure 3.6 Samples of bias (left), thermal (centre), and flat-field (right) calibration images.

Solar CCD Imaging

mask is then applied to sharpen the image, because the binomial mask leaves the image with a soft appearance. The radius of the second mask should be approximately one-third the radius of the binomial mask to compensate for the soft appearance of the first mask. Lastly, the image is saved as a squared (aspect-corrected) 8 bit TIFF file which can easily be imported into other image processing or desktop publishing packages. The script I use is listed below:

LOAD	load raw image
SUBT BIAS.CCD	master bias frame correction
SUBT THERM.CCD	master thermal frame correction
FLAT FLAT.CCD	master flat-field correction
SAVE CCD	save calibrated image, Lynxx format
BUMA 6 6 6	binomial unsharp mask – x, y radius-contrast
USMA 2 2 2	unsharp mask – x, y radius-contrast
SAVE SQT	save squared TIFF image
NEXT	go to next image in list

I can specify the directories from which BatchPIX loads raw and calibration files and where to save images. Using this program, I can start processing images and leave the computer to do the work. The best selection of the day's work can be found by browsing through the processed images.

After BatchPIX has processed the list of images, I select the best and use Aldus PhotoStyler to alter the brightness and contrast scaling, or run a sharpening routine to give the image a final touch. This allows me to make slight adjustments to each individual image to produce the best results. I can also assign a false colour palette to display photosphere with a more "natural" yellow appearance (see Colour Gallery, p. 3).

What features are visible in the solar images? Changes in sunspot structure can be seen over the course of twenty-four hours or less. Imaging sunspots each day will demonstrate the dynamics of the visible solar surface, and these images can be replayed as a movie with the appropriate software.

The focal length of the telescope in conjunction with the pixel size of the CCD camera plays an important

The Art and Science of CCD Astronomy

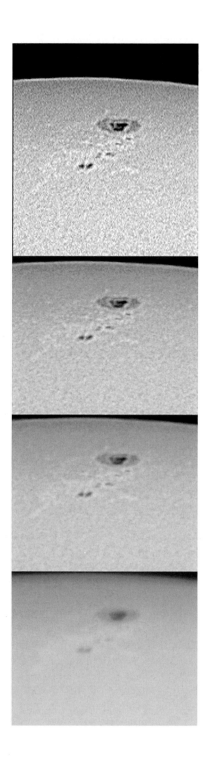

Figure 3.7 Image samples from the different script commands. Far left: original calibrated image. Centre left: BUMA. Centre right: USMA. Far right: aspect corrected and sharpened. 21 May 1995.

Solar CCD Imaging

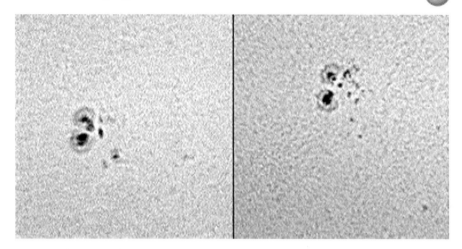

Figure 3.8
Changes in sunspot structure over 24 hours. *Left:* 11 November 94. *Right:* 12 November 1994.

role in the ability to record fine details in solar images. The resolution of the optical system should yield an image scale of approximately 1 arc second per pixel at the CCD. Shorter focal lengths yield a larger field of view, but less resolution. Longer focal lengths increase the resolution, but at the cost of longer exposures and more influence from unsteady seeing. Each observer will find their own optimum setup that will match their telescope, CCD camera and observing site (Figure 3.9).

Details in the umbra and penumbra of sunspots, including light bridges, are easily visible. The shape of the magnetic fields surrounding sunspots can be determined from the shape of the penumbra. Solar pores, sunspots without a penumbra, are visible as well. CCD images display the "Wilson effect", a geometric foreshortening of the spot near the solar limb. Photospheric faculae are bright arcs visible near the limb of the Sun. These are large-scale disturbances which are several hundred degrees hotter than the surrounding photosphere, and mark the appearance or disappearance of sunspots. Intricate details are visible in the faculae as they develop over a period of weeks. Solar granules are also visible, but they require the best seeing conditions to resolve them, because they are at the resolution limit of my CCD and telescope combination. Typical granules range from 1 to 1.5 arc seconds in diameter, appearing as a mottled grey pattern in the CCD images.

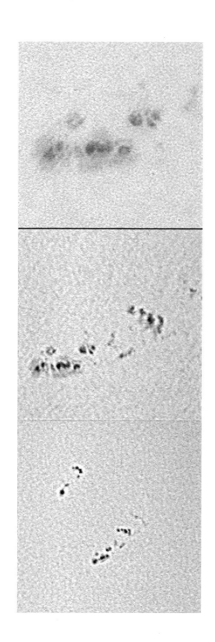

Figure 3.9 Effect of image scale. Left: group of sunspots captured with a 10 cm f/10 refractor. Middle: larger of the two groups with a 25 cm f/10 Schmidt–Cassegrain. Right: same group with 25 cm telescope at f/20. 15 April 1995.

Solar CCD Imaging

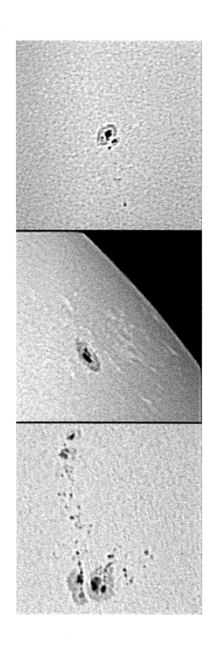

Figure 3.10 Left: large sunspot group showing umbra and penumbra details and solar pores on 15 October 1994. Middle: sunspot near solar limb showing Wilson effect, light bridge, and numerous faculae on 30 October 1994. Right: solar granules and complex sunspot on 7 September 1994.

Conclusions

Electronic imaging is a new addition to the traditional visual and photographic recording methods used in solar observing and record keeping. Readily available computers and CCD technology are powerful tools for amateur astronomers. This equipment is easily adapted to monitoring solar activity, and high-resolution images of active solar regions are within the reach of anyone who has an interest in studying our nearest star. I would encourage anyone with an interest in monitoring our nearest star to invest some time in this fascinating branch of our hobby. I'm sure that you will enjoy the daytime sights while waiting for darkness to fall.

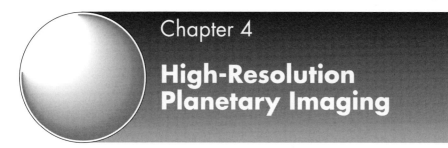

Chapter 4
High-Resolution Planetary Imaging

Gregory A. Terrance

Introduction

Located in upstate New York, Bear Observatory (named after my father) is situated 5 miles south of the small town of Lima. The nearest city, Rochester, is 30 miles to the north, so the skies are fairly dark.

The observatory is 200 feet behind my home and can be operational within minutes. The surrounding area is wooded and grass-covered. Behind the observatory are government-protected wetlands that cannot be used for commercial or residential purposes. All of these conditions promote the steady airflow and good seeing that are needed for high-resolution imaging.

Set-up

The roll-off roof observatory holds four telescopes on a large German equatorial mount. The mount has encoders installed on each axis that are linked to a Meade CAT (computer/database). This makes locating celestial objects fast and easy. All of the Newtonian telescopes are of my construction (Figure 4.1).

Figure 4.1
Gregory A. Terrance with his telescopes in Bear Observatory.

Telescopes

40 cm (16 inch) f/5 Newtonian

This telescope was designed to take high-resolution deep-sky, lunar and planetary images. It's my workhorse. It has two *interchangeable* secondaries. A small secondary is used for high-resolution lunar and planetary work, while a larger secondary is used for deep-sky imaging. The tube has two electrically operated focusers (one per secondary) located 90° from each other. The tube is mounted with rotating rings which allow it to turn for easy access to the eyepiece. This telescope is used for all types of imaging.

25 cm (10 inch) f/6 Newtonian

The 10 inch f/6 was being completed at the time of writing. It has an electrically operated focuser and rotating rings. Its main use will be for lunar and deep-sky imaging.

20 cm (8 inch) f/5 Newtonian

This telescope is used for comets and deep-sky imaging. It also has an electrically operated focuser.

12.7 cm (5 inch) f/10 Schmidt–Cassegrain

Used as a finder, guiding telescope and for imaging.

CCD Cameras

A digital camera is similar to a telescope in that it will favour one type of imaging over another. Large CCD chips are great for imaging big galaxies and nebulae, while the smaller and faster (downloading) chips favour the moon and planets. For that reason, I use more than one telescope and CCD camera.

I still use the original CCD camera that I first purchased in 1992, the Lynxx PC CCD. Although it can take exceptional lunar and planetary images, it has a very small CCD array and fairs poorly when used for deep-sky imaging. Its usefulness may be coming to an end, though, as I have just finished building a digital camera based on the Kodak KAF400 chip. This chip is physically larger than the Lynxx and has 9×9 μm pixels, while the Lynxx has 13×16 μm pixels. If the "shutter-to-screen" time of this new camera can be reduced, it should outperform the Lynxx on the moon and planets. The camera was made through a college course and is a low-noise design, but it needs some refinements made to the software and hardware before it is "user friendly". It should be useful for all types of imaging: high-resolution lunar, planetary and deep-sky.

Physically, the largest CCD that I currently have is the SBIG ST6. I use that and a CFW8 (colour filter wheel) for tri-colour and monochrome deep-sky work. Although this camera is beginning to show its age, it is still very efficient and does a good job on galaxies and nebulae. Finally, I employ a SBIG ST4 for guiding the larger telescopes on comets and deep-sky objects.

Computers

The observatory has a small 4 ft × 11 ft (1.2 m × 3.4 m) heated room where images are acquired and saved on a 486-40 IBM clone. This PC is networked with a computer in the house, a 486-133 IBM clone.

Acquiring High-Resolution Planetary Images

The planetary images on these pages were taken with a 40 cm (16 inch) *f*/5 Newtonian and a SpectraSource Lynxx PC CCD camera. While I have used each of the Newtonian telescopes at Bear Observatory for taking planetary images, the best results have come from the 40 cm because of its greater aperture and resolution.

The steadiness of the area surrounding the telescope will play a large role in how sharp your planetary images are. The steadier the seeing, the better your results will be. It's advisable to locate your telescope away from paved areas (such as parking lots) and rooftops if possible. Large bodies of water and wooded areas quickly shed the heat built up during the day and can be considered elements of a favoured location.

I begin each imaging session by rolling back the roof of the observatory to allow the optics to cool down to ambient temperature. All of the telescopes in the observatory have full-thickness mirrors, which require an hour or longer to cool off. I also check the telescopes' optics to make sure that everything is accurately aligned (collimation). Good collimation is mandatory for getting high-resolution images.

You will need to increase your telescope's focal length if your goal is to image and capture fine planetary details. Eyepiece projection (also called positive projection) increases the focal length of a telescope by placing an eyepiece between the camera and the telescope. The image coming from the telescope passes through the eyepiece and is magnified and projected onto the CCD camera. A camera-to-telescope adapter is required to hold the eyepiece in place. These ready-made adapters are available from any telescope accessory dealer (Figures 4.2 and 4.3).

High-Resolution Planetary Imaging

Figure 4.2 A well-worn eyepiece projection adapter.

Figure 4.3 Focuser, adapter and CCD camera.

The effective focal length of the telescope–projection system will depend on the eyepiece employed and its distance from the CCD chip. In my particular setup, a 16 mm eyepiece is used to attain a 5× increase in focal length ($f/25$) and a 10 mm (0.4 inch) eyepiece is used for an 8× boost ($f/40$). Nearly all amateur telescopes require some type of projection system to make a planetary disk large enough to show surface markings when imaged.

An illuminated guiding eyepiece is an accessory that will come in handy when imaging the planets. These eyepieces are generally used to guide a telescope for long exposures. I use it to centre the planet before placing the eyepiece projection adapter in the focuser. I focus the CCD camera on a planetary disk using very short exposures to freeze the effects of seeing. Once the planet is located and centred on the CCD, I use my guiding telescope to recentre the planet, should it move from the field of view. Take your time when focusing. It may require 5 or 10 minutes to reach the best possible focus.

I generally image Jupiter and Saturn at $f/25$ and Mars at $f/40$. Typical exposures with the Lynxx camera are as follows: Jupiter = $\frac{1}{3}$ second, Saturn = $\frac{1}{2}$ second and Mars = 1 second. I find that a red filter works well on Mars, and that increases exposures by about 10%. Venus is the brightest of all planets. It is so bright that it will "overexpose" or "bloom" most CCD cameras, even when imaged with the shortest possible exposures. The most effective method of imaging Venus is to "stop down" the telescope (reduce aperture by covering the telescope opening) and use filters. Of course, exposure times for all of the planets will depend on your telescope, CCD camera and sky conditions.

As with any kind of photography, reaching and maintaining focus is critical. This is why I consider a well-made electric focuser mandatory for planetary imaging (Figure 4.4). An electric focuser will allow you to make small, incremental changes in focus without touching the telescope. If you need to upgrade your focuser, there are a number of high-quality, commercially available focuser replacements on the market. If you have a high-quality manual focuser, consider adding a motor to it for electronic focusing.

One factor to consider when deciding on a "planetary CCD camera" is the speed at which your camera downloads images to the screen. (As mentioned earlier, I call it "shutter-to-screen" time.) The Lynxx PC camera can download a full-frame image in less than one

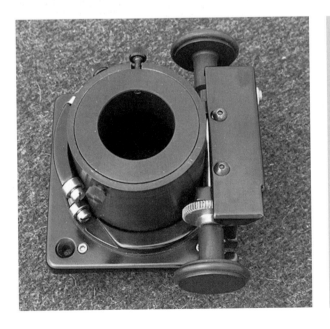

Figure 4.4 A well-made electric focuser.

High-Resolution Planetary Imaging

Figure 4.5 *Top row*: Venus with home-made CCD (16 inch stopped down to 5 inch), two of Mars in 1995 with 16 inch. *Middle row*: Jupiter with 5 inch SCT (note Io, Callisto and Europa); two images of Jupiter with 16 inch showing the Great Red Spot, Io and Europa crossing. *Bottom row*: three images of Saturn with 16 inch in 1992, 1993 and 1995.

second. If an image takes more than 5 or 10 seconds to get from the camera to your monitor, you may find achieving fine focus frustrating as subtle differences in focus are easy to forget after a short time. You might consider purchasing a new or used CCD camera, with a small chip, for planetary imaging. The used market is full of "old" CCD cameras with small chips because their owners have moved on to newer, larger and more costly chips.

One of the advantages of using a CCD camera over film is the ability to see the images as you take them. Use this advantage to its fullest. Take as many images as possible and save what you consider to be the better ones. Although raw images are not as sharp as processed images, you will soon develop a feel for what is a "keeper" and what is not. Unlike deep-sky images, which will often benefit from co-adding multiple images, the planets only require single images to get high-resolution results.

Image Processing

I consider image processing to be an art *and* a science. It's a science because there are standard procedures that must be taken to properly calibrate your raw images. Otherwise noise and unwanted optical defects will be enhanced along with your image. *Bias* and *dark-current* frames must be subtracted from the raw image to remove the noise added by the camera's electronics during an exposure. The remaining image must be *flat-field-corrected* to account for CCD sensitivity variations and optical irregularities. At this point you have a calibrated raw image.

The Lynxx CCD camera is an old design and its images require more calibrating steps than most modern cameras. Images from today's cameras will typically only require flat-field corrections (the bias and dark frames are subtracted automatically). Here are the steps that I take to achieve low-noise, optically corrected images with the Lynxx PC CCD camera:

1. Shoot *raw images* of the planet with an 80% saturation level.
2. Acquire dark frames.
    - Take 10 dark-frame images by covering the telescope and exposing for the same length of time used for the raw images. Later, average these images to form a *master dark frame*.
3. Acquire flat-field images.
    - Take 10 *flat-field images* with an 80% saturation level.
    - Take 10 *flat-field dark frames* by covering the telescope and exposing for the same length of time as used for the *flat-field* images.

High-Resolution Planetary Imaging

- Average the flat-field images. Save as an *averaged flat-field* image.
- Average the flat-field dark frames. Save as an *averaged flat-field dark-frame* image.
- Subtract the averaged flat-field dark frame from the averaged flat-field image to form a *master flat-field image*.

4. Subtract the master dark frame image from the raw images.
5. Flat-field-correct the raw images using the master flat field.
6. The raw images are now calibrated and ready for image processing!

Once you have the noise and optical defects removed from a raw image, it can be saved and led through a myriad of image processing techniques. At this point there are no rules; let your creative side take over. Experiment and try different processing techniques until you find out what is right for a particular image. Never save over the raw image, as it cannot be easily replaced. As you gain experience, your fuzzy-looking raw images will reveal hidden subtle details.

A growing number of image processing packages are available on the market today. Any discussion of a specific software brand or function would soon be

Figure 4.6
Jupiter before processing.

Figure 4.7
Jupiter after processing.

dated. This being the case, I generally try to bring out planetary details without making the planet look false or unnatural. I find that *unsharp masking* tends to enhance details better than *sharpening* does with most software packages. Increasing the contrast will bring out detail. The beauty of the digital darkroom is that you can explore different ways of processing the same image and start over if the results are not what you wanted.

Summary

Today, amateur astronomers have access to hardware and software products that equal or exceed what professional astronomers had a decade ago. While CCD cameras cannot match film's ability to capture wide-field images, there is no question that they excel when turned to smaller objects. CCD images of the planets clearly demonstrate this point. But a CCD's enormous potential can only be realised by taking steps to get the most out of your equipment and location. The following guidelines should get you started in the right direction.

High-Resolution Planetary Imaging

1. Find a favourable location (away from heat sources) for your telescope.
2. Allow your telescope to cool off and reach ambient temperature.
3. Collimate the optics.
4. Use eyepiece projection.
5. Focus carefully.
6. Take many images and save the best.
7. Calibrate images (remove bias and dark current, flat field).
8. Finally - process the best images

Chapter 5
The Comet Watch Program

Tim Puckett

Introduction

The Puckett Observatory, a privately owned facility, is temporarily located about forty miles west of Atlanta, in Villa Rica, Georgia. I started the Comet Watch program in September 1995. The purpose of the program is to monitor activity of all comets brighter than 17th magnitude. Current operations include checking for unexpected magnitude changes and breakups of comets. In December 1995 I began sending precise positions of comets to the Minor Planet Centre in Cambridge, Massachusetts. This chapter will assist you in acquiring CCD images of comets, and offer suggestions on where to contribute your observations.

Setup and Equipment

The observatory building is 16 ft × 20 feet (4.9 m × 6 m), with a roll-off roof, and is of wood construction. The telescope, a 12 inch (30 cm) Meade LX200 Schmidt–Cassegrain, has been modified for complete automation and computer control (Figure 5.1). This has been achieved by replacing the d.c. motors with stepper motors (Figure 5.2). Independent driver electronics control the stepper motors, which are

The Art and Science of CCD Astronomy

Figure 5.1 Tim Puckett's observatory with 12 inch (30 cm) Meade SCT and controlling computer.

Figure 5.2 Stepper motor replacing d.c. motor for computer control.

attached to a 386 computer. The software allows the scope to compensate for proper motion, precession, nutation, annual aberration, atmospheric refraction, and mount flexure. Drive tests confirm that the scope has a total pointing error of ± 15 arc seconds over the entire sky. This pointing error is automatically removed with the PC-TCS software. The software tracks asteroids and comets as well as earth-orbiting satellites by entering orbital elements into the program. It also allows automated drift scans for extended-object spectroscopy. Fully automated observing runs are possible, allowing remote operation from any site with access to a modem.

The scope will soon be accessible via the Internet with a TCP/IP interface using the PC-TCS software. The software to control the telescope's drive system was obtained from Comsoft Inc., located in Tucson, Arizona. The images in this chapter were all taken with an SBIG ST6 CCD camera.

Techniques for Imaging Known Comets

Knowing a comet's position for a given date and time is an essential step in comet imaging. Accurate orbital elements and ephemerides, and other helpful tools for the comet enthusiast, can be obtained from the Smithsonian Astrophysical Observatory (SAO), which operates the IAU's Central Bureau for Astronomical Telegrams (which publishes IAU Circulars regarding discovery and recovery of comets) and the IAU's Minor Planet Centre (which archives and publishes cometary positions in its Minor Planet Circulars). The SAO also publishes the *International Comet Quarterly* (which publishes photometric data on comets, and also issues an annual *Comet Handbook* of elements and ephemerides, primarily for short-period comets).

More information can be obtained from the appropriate SAO offices at the following address (email IAUSUBS@CFA.HARVARD.EDU):

Central Bureau for Astronomical Telegrams/Minor Planet Center/International Comet Quarterly,

Mail Stop 18,
Smithsonian Astrophysical Observatory,
60 Garden Street,
Cambridge, MA 02138
USA.

Or you may visit their Internet site on the World Wide Web at:
http://cfa-www.harvard.edu/cfa/ps/cbat.html

Once the comet elements have been obtained, the next step is to generate an ephemeris. An ephemeris is a listing of predicted positions for a moving object for a given time and date. Many of the programs used today by amateurs for telescope control will automatically generate a comet's position.

I do not use an off-axis guider or star tracker for taking CCD images. A few times a year I spend about two nights getting the telescope polar aligned to within 15 arc seconds of the Pole Star using a 12 mm ($\frac{1}{2}$ inch) eyepiece with illuminated crosshairs. If there is no drift of a star in 20 minutes, the polar error is good enough to begin imaging. Used in conjunction with accurate Periodic Error Correction (PEC) programming, this alignment enables the capture of 5 minute unguided exposures. Exposures of 5 minutes will usually yield a good image of faint comets with the ST6 camera. There are numerous sources available that describe the drift method of polar alignment. Whether one is using a computerised telescope, or an older scope with setting circles, pointing and tracking accuracy will increase dramatically with very precise polar alignment.

To ensure that the telescope has pointed to the correct field, orientate the CCD camera so that images appear north at the top on the monitor. This is also important for obtaining astrometric measurements. I use the THESKY program created by Software Bisque for pointing the scope. This program allows the user to set an aperture box simulating the CCD frame relative to the star map. This is helpful for orientation and scale. After slewing the telescope to the commanded position, I set the focus frame exposure time of the camera to match the Hubble Guide Star projection relative to the THESKY display. The stars in a 2 second exposure match the number and density of stars on the map. Using this method of field verification insures proper field recognition. Faint comets are sometimes very difficult to identify when one is imaging with a CCD camera. One technique is to image a field of a very

The Comet Watch Program

faint comet on two successive nights and then blink the two images to see if the comet appears in the field.

Another technique that I am using is to image while tracking with orbital elements. The modified Meade 12 inch (30 cm) LX200 allows me to track a comet's motion for a given set of accurate orbital elements. Since the photons of light coming from faint, fast-moving comets are scattered across many pixels while tracking at sidereal rate, the ability to record such events will be lost. Keeping the comet centred on the same set of pixels allows me to image fast, faint comets under these conditions.

One reward of imaging comets is to post them on the Internet for other comet enthusiasts to see. A very good place to view current comets on the World Wide Web is The Comet Observation Home Page at the Jet Propulsion Laboratory. This on-line page is run by Charles Morris. Morris is one of the world's best comet observers and keeps the page updated with new images of comets. To view or submit images, visit his page at: http://encke.jpl.nasa.gov/

Observing the changing characteristics of a comet requires imaging on most clear nights. Two periodic comets that have exhibited interesting events are P/Schwassmann–Wachmann 1 and P/Schwassmann–Wachmann 3.

Comet P/Schwassmann–Wachmann 1 has been known to experience one or more outbursts in brightness in a given year. The normal brightness for this comet is around 18th magnitude. These outbursts can brighten it to magnitude 13. Records indicate that the comet has reached magnitude 10 in the past. Figures 5.3 to 5.5 are sequence of this comet taken during early 1996. Figure 5.3 was taken on 25 January at 04:25:04 UT. Note that the comet almost appears stellar. Figure 5.4 was taken on 26 February at 06:37:43 UT. The comet has now brightened and is in outburst. It has begun to eject material. Figure 5.5 was taken on 7 April at 03:20:27 UT. The comet has now started to fade and the material has ejected further away from the comet's nucleus.

Note that the comet images may not be of the same aesthetic quality as some of the other contributions in this book. When comets start showing activity, the imaging conditions may not be ideal for producing a perfect picture. These comet images are intended to be more science-related than beautiful. I have rarely seen research images that look beautiful!

Figure 5.3 Comet P/Schwassmann–Wachmann 1 taken on 25 January 1996.

Figure 5.4 Comet P/Schwassmann–Wachmann 1 taken on 26 February 1996.

The Comet Watch Program

Figure 5.5 Comet P/Schwassmann–Wachmann 1 taken on 7 April 1996.

Figure 5.6 Comet P/Schwassmann–Wachmann 3 taken on 30 November 1995.

Figure 5.7 Comet P/Schwassmann–Wachmann 3 taken on 10 January 1996.

Comet P/Schwassmann-Wachmann 3 in late 1995 and early 1996 is a good example of the breakup of a comet. Here are two examples of this comet before and after the breakup of the nucleus. Figure 5.6 was taken on 30 November 1995 at 23:58:51 UT. The comet appears normal in this first image. Figure 5.7 was taken on 10 January at 01:05:57 UT. The comet has now broken up and the nucleus has split into several components. Since these images were taken with a 12 inch (30 cm) telescope, I was unable to resolve the independent components. However, I was able to record an elongation of the nucleus.

Image Processing

I do very little processing on my comet images. I try to keep the data as clean and undisturbed as possible in case an opportunity arises to contribute the images to professionals. Most of the comets that I image are faint, so changing the brightness and contrast is usually sufficient. One exciting way to process bright comets is with a Rank Order function. This routine was written by author and astronomer Richard Berry. Berry's ST6PIX software has the Rank Order function built into the program. This function allows the user to bring out otherwise hidden details of the jets of bright comets. Here is an example of Comet Hyakutake C/1996B2. Figures 5.8 and 5.9 are the same image. Figure 5.8 is the raw image before using the Rank Order function. Figure 5.9 is the enhanced image showing the jets near the nucleus and demonstrates the power of rank order processing.

Conclusion

The next stage in the development of the Comet Watch program involves several new pieces of equipment. I completed construction of a 24 inch (60 cm) Ritchey–Chrétien telescope in late 1995. This project took ten years to complete. With a focal length of 192 inches (487 cm), the telescope has 16 inch (40 cm) worm gears with oversized worms as speed reducers, and 36.5 inch (93 cm) friction drives on both the right ascension and declination axes. The telescope has a fully computerised drive system.

I purchased three 20 inch (50 cm) Schmidt cameras and plan to start a comet photographic search program. The cameras feature a 5 × 30° field of view and uses 2.25 inch (5.7 cm) roll film. The cameras have large film magazines for film advance and take-up. Each Baker–Nunn camera is capable of photographing 150 square degrees of the sky at 408 arc seconds per millimetre. The photographic magnitude obtainable in four minutes on hyped film is magnitude 18.5. The Baker–Nunn cameras have 31 inch (79 cm) primaries and three-element apochromatic coated correctors. Each Baker–Nunn camera weighs 5 tonnes.

The equipment has been tested and is ready to be moved to a darker location. The 24 inch (60 cm) telescope will be used for CCD imaging and precise astrometric measurements of comets. The Baker–Nunn Schmidt cameras will be used for comet and asteroid discovery. The 12 inch (30 cm) telescope will be used for automated supernova patrol work.

Figure 5.8   Comet Hyakutake C/1996B2; raw image.

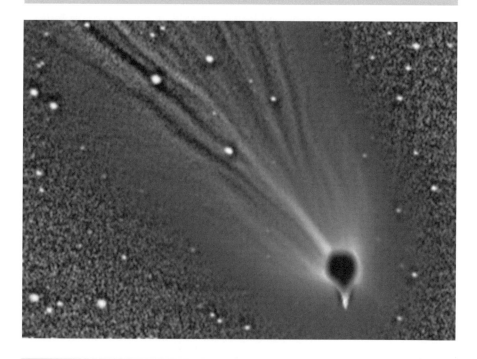

Figure 5.9   Comet Hyakutake C/1996B2; processed image showing the jets near the nucleus and demonstrating the power of rank order processing.

The Comet Watch Program

You can visit the observatory on-line through the World Wide Web at:
http://www.mindspring.com/~tpuckett

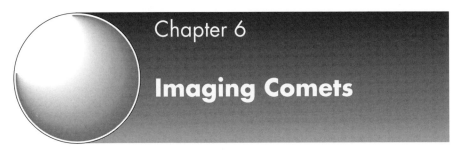

Chapter 6
Imaging Comets

David Strange

Introduction

Worth Hill Observatory is situated on the South Dorset coast in the UK and is fortunate to be in a rural area, with dark skies and a clear sea horizon from east to west. The telescope is a 500 mm (19.5 inch) $f/4$ Newtonian reflector mounted on a German equatorial and is housed in a 4 m (13 feet) dome (Figure 6.1).

The observatory is a plastic-coated steel structure and has been designed so that I can observe objects at low declination, since it rotates on a two-foot-high dwarf wall allowing the telescope to view objects close to the horizon (Figure 6.2). This fact enabled me to image Comet Hale–Bopp just six degrees above the horizon (Figure 6.3).

Set up

I have a Starlight Xpress CCD camera (Figure 6.4) and framestore linked to a 386SX computer, which has itself performed very reliably over the past three years, considering the wide ranging conditions of temperature and humidity with which it has to contend. The camera outputs its image to a separate CCTV monitor, which has the advantage of enabling the computer to perform

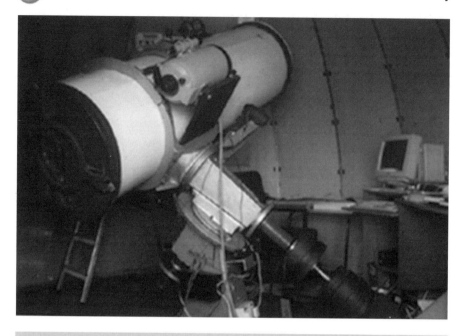

Figure 6.1 David Strange's 500 mm f/4 Newtonian reflector.

Figure 6.2 Worth Hill Observatory.

Imaging Comets

Figure 6.3
Comet Hale–Bopp image.

other tasks such as displaying digital star fields while the camera is capturing images.

Image Acquisition

The telescope has a 6 inch (15 cm) guide scope attached. Although I have often used this for guiding 160 second exposures, I tend now to make a short series of 40 second exposures and co-add them, which circumvents the periodic errors of the drive system.

A standard system which I have now adopted is to take three to four images together with a dark frame and flat field on each 1.44 Mb floppy disk. These images are then processed on a faster 486DX 66 MHz computer in my office. 4×40 second exposures generally reach to magnitude eighteen with the 500 mm $f/4$ and CCD. The field of view obtained in the image with this configuration is 10×7 arc minutes. Wider fields can be obtained by using 135 mm lens (2.5° field) or 28 mm lens ($12 \times 8°$ field).

Focusing has to be quite critical to get sharp images, and I have found that an electric focuser is a great boon for making fine adjustments when scrutinising stellar images on the monitor.

Figure 6.4
Starlight Xpress CCD camera and electric focuser.

Imaging Comets

Comets are perhaps the most intriguing of celestial targets to capture on CCD images. Their nightly changes of appearance offer fascinating opportunities to the CCD imager. We can capture short-lived disconnection events in the tail, or record plumes or jets from the comet nucleus.

There are several aspects of comet imaging that test one's telescope and CCD equipment to the full. For a start, the comet may be so faint that it may prove difficult to locate in the first place. This was the situation when I tried to find the faint comet train of Shoemaker–Levy 9. With each of the comet fragments being in the order of 20th magnitude they were beyond the visible reaches of my 500 mm (20 inch) telescope.

Imaging Comets

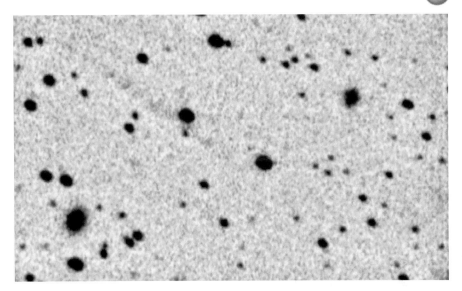

Figure 6.5
Comet Shoemaker–Levy 9.

In such cases you need an accurate ephemeris and be able to plot the comet's position with respect to the surrounding stars as shown by a digital star atlas such as the *Hubble Guide Star Catalogue*. Only by centring and imaging the surrounding field stars is it thus possible to locate the comet. This technique I have christened "digital star hopping". The image above is a 5 × 160 second exposure which has only just begun to record the comet. This image has been highly contrast stretched, so there is much background noise (Figure 6.5).

Second, the coma or dust cloud of the comet is generally much fainter than the nuclear region or central condensation. It is therefore not easy to enhance the faint detail in the coma without saturating the nucleus. The best routine to use here is a logarithmic stretch, which will tend to compress the brighter regions of the image but expand the fainter and darker areas.

One feature of comets that remains the most elusive is that of intricate tail structures. Not only can they remain stubbornly faint, but they can also appear impressively large – orders of several or tens of degrees are not uncommon. Bearing in mind the small field of view of CCD cameras, several different routines are called for to capture long tails on CCD images! If your only option remains with using a telescope with a focal length of 1000 to 2000 mm then your best choice is to construct a mosaic of two or three images spanning the length of the comet's tail. This is what I achieved

Figure 6.6
Comet de Vico.

when I imaged Comet de Vico in September 1995; the result is a mosaic of three 10 × 7 arc minute images (Figure 6.6).

One needs to make sure that each image overlaps by a few arc minutes to permit accurate registration and alignment. However, bear in mind that the comet is likely to move between the first and last image taken. This means that there will be a noticeable movement against the background stars, so do not use the stars to align the comet or else the tail will not join up.

The software which I use to construct CCD mosaics is Paintshop Pro 3.11, which is a very useful general purpose image processing program. I tend to use Pixwin for the initial stages, but rely on Paintshop Pro for adding text and manipulating images. One first has to open a new image or background canvas onto which the comet mosaic will be pasted. I would suggest making this sufficiently large, say 1024 × 800 pixels in size, so that we can paste three comet images side by side, each of which needs to be resized to 320 × 200 pixels. The overall effect is achieved by copying and pasting the original images in sequence on the background canvas, and in addition there are various tools available in the program to make a smooth, joint-free connection between one image and the next.

A slightly easier option for recording long tails is to use shorter focal lengths, and to this end I use 135 mm or 28 mm camera lenses. A fast $f/2.8$ lens can record an impressive amount of tail, as the image of Comet Hyakutake 1996 B2 shows (Figure 6.7). Such a lens gives us a field of view of 12 × 8 °.

Imaging Comets

Figure 6.7
Comet Hyakutake.

Beware of "false comets" when using camera lenses and CCDs. Any star emitting an infrared excess in its radiation will appear out of focus and look vaguely like a comet. If you find any of these "bloated stars" try refocusing a little and you will find this star will now sharpen up while all the others will lose focus.

The central condensation of comets provides yet a further interesting area to explore with the CCD. This is particularly so with bright comets such as Swift–Tuttle and indeed our most recent visitor Comet Hyakutake. This particular comet was moving so fast that I restricted the exposure to 6 seconds when imaging through the main telescope. This was sufficient, however, to record an impressive amount of detail in the bright nuclear region. The raw images displayed dark nuclear shadows, a main spine-like jet aligned on the tail and numerous curving jets on the sunward side. From a record of the images obtained on the nights of March 27/28 1996 it was possible to determine the 6.25 hour rotation period of the comet by viewing the changing orientation of the nuclear jets.

The images of Comet Hyakutake (Figure 6.8) show just how much information we can glean from our CCD images. The left-hand image is a raw 6 second exposure which has been dark-framed. The central one has been given a gentle unsharp mask, while the image on the right has been given a high-pass, high-power filter.

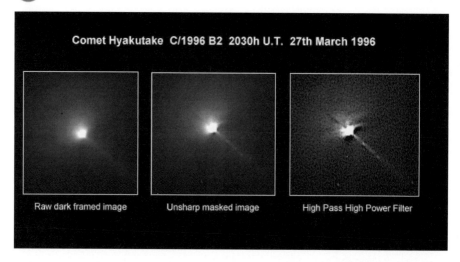

Figure 6.8
Hyakutake's nuclear jets: effects of image processing.

Near-nucleus studies of bright comets allow us to build up a useful library of images showing the rapidly changing orientation of the comet's jets. We can now go on and link these images together to make a computer animated sequence or CCD flick. Your images take on a whole new dimension when showing a rotating comet nucleus spewing out material like a cosmic garden sprinkler!

Chapter 7
Nebulae and Galaxies in High Resolution

Luc Vanhoeck

Introduction

For the last twenty-five years or so I have been looking at and photographing the heavens. It started with Comet Kohoutek. Not that I ever saw it. It was just too faint and I was probably too inexperienced. But it awoke my interest in astronomy, and since then the stars have really become part of my life. There are few things astronomical I have not been involved in – an astrophotography section, a magazine, astronomical camps and popularisation projects. From 1989 to 1994 I served as editor-in-chief of *Heelal*, the magazine of the Flemish Astronomical Association. The first years I was collaborating with Ignace Naudts. After his death an asteroid was named after him. For the past few years I have concentrated on CCD imaging. My friend Staf Geens, himself an avid astrophotographer, has helped me a lot in this, adapting my telescope over and over again to assure the best possible images.

The CCD story

Right from the first advertisement for the ST4 camera in *Sky and Telescope*, I knew it: this was the beginning of a new era and I wanted to be part of it. A couple of weeks later I was experimenting with this camera. In 1993 the adventure really began. With the aid of Hugo Ruland, a local manufacturer of astronomical telescopes, I learned about the ST6. It was coupled to an 8 inch (20 cm) LX200 and I obtained hundreds of images with this setup.

Early 1995 we decided to upgrade to the ST8 camera, equipped with the 1536 × 1024 pixels Kodak chip. At once, I was convinced that the real potential of this CCD could only be demonstrated using a larger telescope. I opted for a Celestron CG11, mainly because of the very precise mounting (Figure 7.1). At the same time it was a surprise indeed what a difference an 11 inch (27.5 cm) can make!

Figure 7.1 The CG11 SCT in Luc Vanhoeck's observatory.

Nebulae and Galaxies in High Resolution

Figure 7.2 The skylight extends through the sloping roof of the house. This may not be a textbook example of an ideal observatory, but it offers the observer some comfort.

We happened to build a new house in 1990 and I chose a skylight that extends through its sloping roof. This flips up quite easily, so that the telescope is ready for use within a couple of minutes (Figure 7.2). The skylight measures only 1.5 × 1.5 m (5 by 5 feet), so an 11 inch (27.5 cm) SCT is absolutely the largest telescope it can accommodate.

Next to the observatory I have a small study. It not only contains the computer equipment but also star atlases and magazines that I consult quite often, always looking for new and challenging objects to photograph. Our house is situated between Antwerp and Brussels, possibly the most heavily light-polluted area on the continent, apart from the city centres. Belgium has a dense highway network that is illuminated during the whole night! I certainly do not have excellent eyes: my limiting magnitude does not exceed magnitude 3.5 to 4 very often. So I was very fortunate that the CCD came just in time, or this situation might have changed my astronomical activities negatively.

Image Taking Procedure

When taking images, I spend a lot of time focusing the camera. I found this to be the most critical part of CCD imaging and the time taken to do it well pays off later! Usually, I just take a piece of cardboard with two symmetrical holes (Figure 7.3). A star shows up double

Figure 7.3 This simple mask guarantees sharp images all the time.

unless you reach the focal point. This way it is possible to obtain correct focus repeatedly.

It is very important that the telescope and the CCD camera have about the same angular resolution. I use my ST8 at medium resolution, i.e. with 18 μm pixels that cover 1.3 arc seconds each. Working with smaller pixels does not make any sense, because the telescope is not able to show smaller details in a 15 minute exposure. In addition, the file size grows unacceptably.

The Kodak chips are inherently less sensitive than e.g. the Texas Instruments chip that is used in the ST6 camera. This may lead to 10 times the exposure time, but by combining the pixels two by two only 4 times the exposure time is needed. Most of my images are exposed for 15 minutes and sometimes even more, depending on the sky conditions. This long exposure time

Nebulae and Galaxies in High Resolution

pays off too. It is possible to reach 14th magnitude stars in only 10 seconds, but the signal-to-noise ratio improves drastically the longer you expose. Most of my images have been processed using maximum entropy deconvolution or an unsharp mask. Both techniques benefit from well-exposed images. The only problem is that you find yourself with exposure times comparable to classical astrophotography, a technique you abandoned partly because you thought CCDs would offer you much quicker results.

Another myth about CCDs is that it doesn't make any difference if they are used under severe light pollution. There is no substitute for dark skies: the better your sky condition, the better your results. Let nobody tell you anything else!

The M16–M17 Duo in the Southern Milky Way

Two of my favourite nebulae are M16 and M17 (Figures 7.4 and 7.5 respectively) in the southern Milky Way. I obtained the images of them in France where I travel

Figure 7.4 The beautiful nebula M16 Serpentis required an exposure time of 15 minutes.

Figure 7.5 This picture of M17 Sagittarii was processed with an unsharp mask to enhance the subtle details; the exposure time was also 15 minutes at f/10.

yearly with a local group of amateurs to escape the local light pollution.

We arrived there in late August 1995, and after a few rainy days the sky cleared. The group went out to dinner and I had to bring my wife and three daughters to the vacation house we had rented high in the mountains. It was extremely dark and a very difficult drive. It took me about 30 minutes to drive these 8 kilometres and again 40 minutes to return to my friends. Needless to say I was quite tired after this dangerous mountain drive and it took me more time than usual to set up the telescope. Imagine the night that followed. It was very dark and the brightest deep-sky objects were visible to the unaided eye. But I really was not able to image a single one. Nothing worked, everything went wrong, and in the end I was not even able to centre any object on the ST8 chip. I finally gave up and waited till dawn to drive to our house, with lots of empty floppies. I awoke a few hours later to discover symptoms of a disease that will require changes to my usual life style. It was definitely the most frustrating experience in more than twenty years' stargazing.

That day I brought my telescope to our house and set it up next to a steep abyss. When it became dark, I did not even dare to walk around it, so scared was I of falling fifty metres down! On the other hand, I was determined to make my best images ever. And I did. As if I had a textbook in my head, I checked about every possible

Nebulae and Galaxies in High Resolution

detail. The polar alignment was perfect, focus was perfect and guiding was perfect – I just did not allow anything to go wrong. Failure was not an option, as they used to say, though it took my utmost concentration.

Other Nebulae

Recently, I put all my ST6 and ST8 images on a CD because I was running out of floppy disks! I have hundreds of images and it's difficult to choose from them for this chapter. Anyway, the Crab Nebula is one of my favourites because it shows details I could just dream about only two years ago. The image (Figure 7.6) was taken from my backyard observatory. The sky was very clear but it was already becoming late. Nevertheless, I resisted the temptation to apply a shorter exposure time and exposed for 15 minutes as usual.

Galaxies

CCD imaging can be a strange process. Sometime in January 1996 I tried to make pictures of the Moon. In

Figure 7.6 M1. It is hard to believe this picture was taken from a heavily light-polluted site!

Figure 7.7 A compilation of galaxies, most of them photographed using a red filter.

order to dim it a little bit, I put a red filter in the light path. Afterwards I forgot it, only to discover it after taking images of the nice galaxies in the spring skies. Using a red filter accentuates fine details in such objects and at the same time moderates the influence of heavy light pollution. Just by accident, I found a way to take better images! The picture (Figure 7.7) shows NGC7479, NGC891, M82, NGC2903, M66, NGC660, M65, M51, M95 and NGC7814, each exposed for 15 minutes at $f/10$.

Nebulae and Galaxies in High Resolution

Processing Images

By trial and error I found it to be extremely important to use all the information the image contains. The ST8 delivers 65 536 grey shades, and at first this seems to be an exaggeration. Computer screens can only show 64 of them and our eye just discerns up to 40. Early software packages transformed the images rather quickly to 256 colour versions. This is not the best option. Image processing is able to accentuate minuscule brightness differences. It does not make any sense, however, to throw away such information before making use of it. At our office, we use Photoshop 3.0, which has a very powerful algorithm to redistribute all the grey shades in an image over the visible range. Using this technique is the best way to give the image the necessary power!

Thanks to my friend Rik Blondeel, I discovered Hidden Image software, a package that has an extremely powerful maximum entropy deconvolution algorithm. This is an iterative method, meaning that it works in steps, with each step a little bit better. Noise, attributable to the electronics but especially seeing effects in the atmosphere, prevents you from obtaining perfect, diffraction-limited images. But in general, I can improve the resolution at least twofold! My own 33 MHz 486 computer is too slow to run Hidden Image (a single ST8 file may take a night to process), so Rik helps me with his faster PC. In fact, these are my trade secrets: applying an exposure time that is comparable to traditional astrophotography, a red filter most of the time now, making good use of all the grey shades the camera can deliver and applying unsharp masks or maximum entropy deconvolution where appropriate (Figure 7.8). Nothing more, nothing less.

Personally I am convinced that every dedicated amateur is able to obtain nice pictures. It is true that the necessary equipment is still quite expensive. This may change, I hope. On the other hand, getting such images may take time. I feel it to be important to have the necessary experience as a traditional astrophotographer. One needs to know how to focus a telescope and how to guide it well. A second prerequisite is that you know the principles of working with a PC in general and image processing in particular. I have been working with computers from the early 1980s, and this knowledge helped me to a large extent.

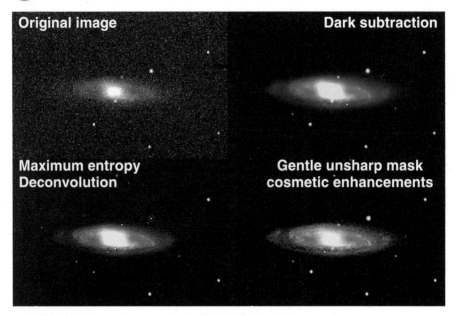

Figure 7.8 The various steps in processing an image (see text).

Colour Images

The same principles as described above hold true for my colour images. There is one additional factor, however. Individual black-and-white files have to be combined into colour. My first attempts failed because I went to 256 colour versions of the green, blue and red files too quickly. I always keep the original files, however (this can take lots of floppy disks), and later on I discovered that much better results can be obtained by keeping the 16 bits or more than 65 000 colours as long as possible. In fact, this turned out to be even more important than with just black-and-white images.

A second problem was finding out the "best" colour for each object. My colour images were taken in the prime focus of a 4 inch (10 cm) $f/5$ Genesis refractor (Figure 7.9). The filtered green, blue and red ones were exposed for 10 minutes each, and this does not correspond with the filter factors one should apply.

I compared and sometimes adapted my results to published images, if possible made by the well-known British professional astrophotographer David Malin. But personal taste played an important role too. Of course, I am the first to admit my colour images cannot

Nebulae and Galaxies in High Resolution

Figure 7.9 Luc Vanhoeck, seen here photographing a partial solar eclipse, uses a 4 inch f/5 to take colour images.

be used for any scientific analysis. That is not why I am making them. I just like to do so, without pretending anything else. Examples of my colour images are included in the Colour Gallery (p. 7).

Why?

One can indeed ask why one makes CCD images just for purely aesthetic reasons. It may sound hard to some, but I like it and I have no other driving forces! I know my equipment can be used for more scientific purposes, but I dislike the feeling that I need to do systematic observations. I definitely need variation. That's why I no longer make any commitment. On the other hand, I work with my CCD and computer with fanatical intensity. I will never feel comfortable with less than perfect images. I am always looking to improve on them, but a really good image can only make me feel good for a couple of days. Thereafter I need to make another one.

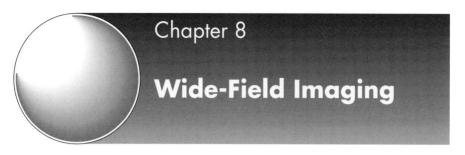

Chapter 8
Wide-Field Imaging

John Sanford

Introduction

After about forty years of astrophotography, the CCD revolution hit me hard. Here was a way to make images almost instantly and without guiding, a chore I had come to regard almost the same as a trip to the dentist! I first purchased an ST4 in order to auto-guide, but did some imaging with it to see what I could do. The results were interesting but seemed primitive, and when I heard about the Starlight Xpress I thought maybe here was something that would give me an almost photographic-looking 5 × 7 inch (12.5 × 17.5 cm) image. I also decided to become the US dealer for the factory in England, since they were looking for a US outlet. The business aspect has been less than a huge success, but there are over twenty cameras in use here, in Mexico, and in Alaska. The camera has been used by many workers around the world, and is a favourite in England and Italy, among other countries. Supernovae and asteroids have been discovered with this camera from Italy and England, respectively.

My Observing

I observe with my own Celestron 14, using either an *f*/7 Celestron focal reducer-coma corrector, or the more

Figure 8.1
Orange County Astronomers' Observatory at Anza, California. John Sanford's C-14 is seen at right

recent Optec MaxField focal reducer, which brings the C14's $f/11$ focal ratio down to about $f/4$. Also, for high-resolution lunar and planetary work, I enlarge the image somewhat with a 2× or 3× negative projection lens. I have also used an 8 inch (20 cm) LX200 ($f/10$) loaned to me by Meade Instruments Corporation, and my own 12 inch (30 cm) $f/10$ Ritchey-Chretien, which I have set up in my backyard here in Costa Mesa, California.

I am lucky to have the C14 on a permanent pier at the Orange County Astronomers' Observatory at Anza, California. This is a dark-sky site about 95 miles (150 km) from home, near Palomar Mountain, at an altitude of 4350 feet (1325 m). I try to go to the observatory once a month, and generally stay up almost all night when there. Anza has about 300 days of sunshine during the year, and the coldest temperatures in winter are about freezing in December, January, and February.

The CCD Camera I Use

For about four years, I have been using the framestore model of the Starlight, which has a 510 × 256 pixel chip made by Sony. The data come from the camera and go into a framestore memory, which also has a video

Wide-Field Imaging

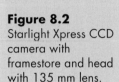

Figure 8.2 Starlight Xpress CCD camera with framestore and head with 135 mm lens.

output, so the picture becomes visible immediately after taking and downloading off the chip.

This is one of the features of this camera, because in planetary, lunar, and solar imaging you can quickly examine each image, as the camera repeatedly takes an image, and delivers a new one every 3 seconds or so to the monitor. When a particularly good image appears, it can be saved by flipping a switch. Thus only the sharpest images are kept, thereby saving memory space and assuring good results. This is very convenient, because the atmosphere blurs a good percentage of exposures during any observing session except on the very rare perfectly still night. Even some longer exposures (say "integrations") of deep-sky objects need to be discarded because of wind jiggle, bad tracking, or bad focus. The framestore allows editing without any need to save to the computer memory. I also use the video image on a 20 inch (50 cm) monitor at public events when saving and processing the image are not necessary.

Image Processing

One of the keys to making good CCD images that are publishable, however, is the processing that can be done to a raw image. Digital processing has far more possibilities than photographic, allowing stretching of contrast various ways, unsharp masking to reveal finer details in lunar and planetary images, and various forms of maximum entropy deconvolution, which pull out the most detail in galaxy images, for example. Elsewhere in this book, it will be noted that the CCD imager almost always makes dark frames and flat-field frames that are only slightly exposed in order to subtract noise and uneven-gain pixels, as well as any dust or vignetting that might occur in the optical system. Care to obtain as much precision as practical at this stage will result in images that are as good as possible when completely processed. This pre- and post-exposure work is generally called "calibration", and will result in scientifically accurate images if performed correctly.

Wide-Field Work

One area of CCD imaging is wide-field imaging. The definition here might be any work using a focal length of less than 500 mm or so. This will be largely "piggyback" work, where the camera and lens are mounted on a driven telescope, or even on a "camera tracker" type of mount, essentially a miniature equatorial mount made to carry a small camera, not a telescope. Owing to the tiny size of a CCD chip (about 4×6 mm), there will be *a focal length magnifying effect* that is astounding to 35 mm film photographers. The CCD camera with a miniature chip and 135 mm lens has a field of only about 2 degrees – the same as a 1000 mm focal length telescope with 35 mm film!

Really wide-field imaging is very difficult with the CCD chip camera, as very short focal lengths (<10 mm)

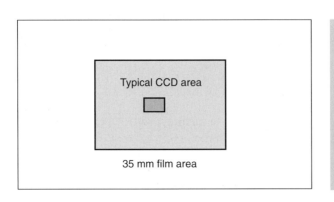

Figure 8.3
Comparison of CCD chip size and 35 mm film area (hence field on sky).

Wide-Field Imaging

are needed to achieve what a 28 mm would capture on 35 mm film. The use of cinema or video lenses would be helpful here, but their short back focus (distance from lens to film) is so short that most cooled CCD cameras with a glass window have too much space inside the window, precluding the placement of a cinema lens near the chip.

Lenses

As a practical matter, then, one is limited to the use of 35 mm camera lenses, because they have enough room behind the lens to mount on most CCD cameras. The Starlight Xpress has a 42 mm screw thread moulded into its head casting, so that fairly common Pentax–Praktica screw thread lenses can be used directly. I have been able to find three such lenses for my work without too much trouble, at a local "photo swap meet", all at reasonable prices. I have used a 28 mm $f/2.8$, a 50 mm $f/1.4$, and a 135 mm $f/2.5$ Super Takumar to make the accompanying images. Generally I stop the lenses down at least half a stop from wide open as a matter of habit, to improve definition and contrast, although this practice isn't strictly necessary with CCD work, as we are using only the centre of the 36 × 24 mm field with the 4 × 6 mm chip.

"Exposure" and Definition Problems

When exposing (integrating) with a small lens at a fairly wide focal ratio such as $f/3.5$, one reaches the sky fog limit quickly if no filter is used. A few minutes is as long as is necessary for most work, and will show, for example, the Milky Way star clouds very nicely (Figure 8.4b). I have found that the use of a coloured filter to limit the bandwidth of light to, say, green, yellow, or red will make the stars sharper and smaller with my CCD camera. The reason is probably that the CCD chip is sensitive well into the infrared (IR), and many stars have quite a lot of their radiation there. If you focus the camera on the visual rays of such a star, then the IR

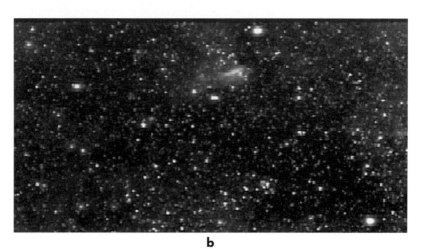

Figure 8.4
a Field near M37 unfiltered, showing red star. **b** Field with M17, made with green filter showing smaller stars.

excess will cause a halo around the star image, since the IR light hasn't reached a focus yet. This form of chromatic aberration is a serious problem in CCD wide-field astroimaging. There are some stars whose radiation is largely infrared, and these stars appear as fuzzy disks or cometary on my images (Figure 8.4a).

Only the use of an infrared rejection or blocking filter will alleviate this problem. To date I have not employed such a filter, as they are expensive and difficult to find in a round 48 or 52 mm diameter to cover the 135 and 50 mm lens apertures. I have relied on the green or red filters and image processing to shrink the size of my star disks to acceptable diameters.

Wide-Field Imaging

Image Processing Tips for Wide Field Pictures

The image processing here is little different from other CCD procedures. Dark frames for hot pixel removal should be made at the working temperature and times, as well as several twilight out-of-focus calibration images ("flat frames") for vignetting and dust image removal during processing. Milky Way frames will be well exposed, so some midrange contrast stretching might be attempted that isn't often employed with deep-sky images. Also, the "high-pass" filter can be used with such frames to make the star images almost photographically small. However, there is an increase in the background graininess, so the final result looks like it was shot on T-Max 3200 film or worse.

Often more information may be seen in a negative image. This is the way most astronomers like their images, since faint and delicate detail may be lost in the black of a normal or positive image (Figure 8.5).

Figure 8.5 Virgo cluster of galaxies, centre area (M84 and M86 right centre). 135 mm lens at f/3; 70 seconds integration, Starlight Xpress CCD camera.

Conclusion

Wide-field imaging may be used for several purposes. The CCD camera allows penetration of space in short times so that nova, comet, and supernova patrol work could be carried out easily to 14th magnitude with a 135 mm lens. Even so, owing to the small field, many exposures must be made to cover the area of one 35 mm film negative. Whether this is time-effective remains to be seen at this early date in the employment of amateur CCD cameras. Certainly photometry may be carried out on wide field images, and the study of faint variable stars with amateur CCDs is a field that will no doubt develop in the near future. The recording of meteor shower activity is another area where wide-field lenses could produce excellent results, provided fields of larger than 45° are attained.

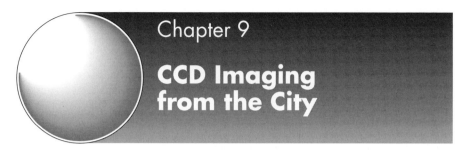

Chapter 9
CCD Imaging from the City

Adrian Catterall

Introduction

I have been interested in Astronomy for twenty-five years and bought my first telescope as a teenager. After a few years of satisfying visual observing my career took me to London in 1978 where I have resided ever since. Because light pollution filters were not widely available and, because I lived in a succession of gardenless flats, my interest waned. In 1989 I invested in a then widely available light pollution filter and bought accessories necessary for astrophotography which rekindled my interest with some initial success.

However, to achieve more satisfying results I often travelled ninety miles to darker skies, which was a huge effort considering I had an 11 inch Celestron SCT and accessories coupled with the unpredictability of England's weather. My interest could easily have declined but for the timely introduction of amateur charge-coupled device (CCD) cameras in 1990.

My current equipment includes a 12 inch Meade LX200 and two Santa Barbara Instrument Group cameras, ST6 and ST7 (Figure 9.1). I am located five miles south of the centre of London, where there is considerable light pollution with a naked-eye limiting magnitude of about 3.0. I have found visual observing of the planets and moon rewarding, but deep-sky observing is virtually impossible with only the Orion Nebula and a few bright galaxies offering some

Figure 9.1
Adrian Caterall and his 30 cm Meade LX200 with ST7 CCD camera.

satisfaction. In addition to a broadband light pollution filter I use two narrow-band filters, the Oxygen III and UHC filters which considerably enhance contrast of certain nebulae such as the Dumbell and Veil.

CCD cameras were introduced to the amateur market in 1990 and were heralded as ultra-sensitive, revealing detail in the spiral arms of galaxies with short exposures even from light-polluted skies or with a nearby full Moon. Well, this was an opportunity I could not resist, so my CCD imaging career began in 1992. I was initially frustrated by the demands of a CCD; to obtain good images the magnification has to be higher and the field of view smaller than in conventional photography film, and both an accurate drive and good polar alignment are essential prerequisites. With the introduction of larger CCD chips and autoguiding CCD chips, periodic error correction on commercial telescopes, centring objects and tracking them became easier.

Methods of Overcoming Light Pollution

Image at the Zenith

If your conditions are similar to mine then you are likely to be surrounded by other houses and perhaps nearby industrial estates that emit not only light pollution, but air pollutants and heat which degrade image quality through the telescope. As light pollution decreases with increasing altitude and seeing conditions improve away from local heat sources (particularly in the winter) such as chimneys and central heating, higher-quality images can be obtained from objects at or around the zenith. I have also found better seeing conditions in the early morning hours when most residents are asleep and their central heating is switched off! Seeing conditions are particularly relevant to CCDs because cameras with small pixels can achieve high resolution and may oversample stellar images with typical amateur-sized scopes. Therefore seeing conditions become a limiting factor in the final image resolution, which is in contrast to visual observations where transparency is more important to detect faint deep-sky objects. Light pollution will increase the background noise and reduce the final signal-to-noise ratio (see below), hence imaging at the zenith will give you the best overall quality (Figure 9.2).

Longer exposures

Image quality is dependent on the signal-to-noise ratio. The noise of an image can be calculated from reading the pixel counts of that image from the software accompanying the camera. The total noise from an image can be calculated by adding parameters such as readout noise, dark current noise and background noise. Even processing, such as flat fielding and dark-frame

subtraction, produce noise. The total can then be calculated from the formula

$$N = \sqrt{(N_1^2 + N_2^2 + N_3^2 + \cdots)};$$

see *CCD Astronomy*, Summer 1994, pp. 34–8. In light-polluted skies the background will give the largest noise contribution followed by the dark-current noise inherent in the camera. For example, I have saturated the SBIG ST7 camera in five minutes when imaging the Omega Nebula at only 17° above the horizon and after only 30 seconds when I imaged Comet Hyakutake with an ST6 and a 50 mm camera lens (Figure 9.3).

Techniques to overcome the noise from background light pollution include longer exposures which increases the signal-to-noise ratio by a factor of √2 for a doubling of exposure length. If the chip is near saturation then averaging multiple exposures will give similar results (Figure 9.4). I have found that exposures approximately four times longer are required to achieve a similar signal-to-noise ratio as from a dark location. More signal can be achieved by reducing the focal ratio of the telescope; typically I use a Meade f/6.3 focal reducer and have recently acquired an Optec f/3.3 focal reducer which gives four times more light, effectively reducing exposure length. The field of view is also four times larger (28 × 18 arc minutes on my 12 in LX200).

Figure 9.2 The Bubble Nebula (NGC7635) taken near the zenith. The full extent of the nebula is extremely faint, but shows well in this image. Meade 12 inch (30 cm) LX200 at f/3.3 with a Lumicon broad-band light pollution filter, 600 second exposure with an ST7 on 9 December 1995.

CCD Imaging from the City

Figure 9.3
Comet Hyakutake taken with an ST6 attached to a piggybacked Olympus camera and a 50 mm lens yielding a field of view 10 × 8°. This 6 second exposure demonstrates the marked light pollution at the lower right. A 30 second exposure saturated the chip.

The dark current can be reduced by cooling the camera as low as possible, and some newer cameras (e.g. those using the KAF0400 chip) have inherently lower levels. CCD cameras in non-light-polluted skies are said to be background-limited, as the dark-current will build up slowly and allow hour-long exposures without saturation.

Light Pollution Filters

Light pollution filters are aimed at reducing background skyglow, with a consequent increase in the contrast of deep-sky objects; however, they will reduce some of the light from deep-sky objects, and are therefore not a substitute for a good dark-sky location. There are many sources of light pollution, and certain pollutants in particular will leak through filters: for example, common light bulbs emit across most of the visual spectrum. Today's street lights commonly use high-pressure sodium vapour, which shines across a broad spectral range but has peaks in the region of 550 to 630 nm. Popular security lights utilise mercury vapour and emit strongly at 405, at 436 and between 540 and 630 nm. Low-pressure lights emit around 590 nm. Standard broad-band light pollution filters reject light at the central region of the visible spectrum

Figure 9.4 The Horsehead Nebula (IC434) and NGC2023. I used an average of 4 × 300 second exposures because this object was at a low altitude and a single 1200 second exposure would have saturated the CCD camera.

between 550 and 650 nm and at levels below about 450 nm. Therefore most of typical light pollution sources are filtered out except the common light bulb from neighbours' houses; here a friendly, courteous request may prove fruitful. However, certain deep-sky objects such as globular and open clusters and many galaxies emit their light across a broad range of the spectrum, and much of their natural light will also be filtered, but some gain will still be achieved from broad-band filters (Figure 9.5).

Other deep-sky objects emit light in narrow bands; for example, emission nebulae such as the Orion and Lagoon Nebulae emit their light at wavelengths 486.1, 495.9 and 656.3 nm, corresponding to hydrogen-β, oxygen-III and hydrogen-α, respectively. Light from these will mostly be transmitted through broad-band filters, and consequently will have the most contrast gain (Figures 9.6 and 9.7).

More specialised light pollution filters such as the oxygen-III or UHC filters only transmit very narrow bands and are more suitable for visual use on select objects such as the Veil Nebula. Stellar images are greatly suppressed with these filters. To date I have not used these for CCD imaging.

From a light-polluted sky, I suggest using a broad-band filter and selecting objects that emit light in a

CCD Imaging from the City

Figure 9.5
NGC891 in Andromeda from a longer 40 minute exposure taken with a Lumicon broad-band light pollution filter. Meade 12 inch (30 cm) LX200 at f/3.3 on 28 September 1995.

narrow spectrum. They should be imaged when they are at culmination, i.e. at the highest declination possible, and in the early hours of the morning. Where possible, save other objects for a time when you can visit a darker location.

Colour Imaging

Light pollution can play havoc with tri-colour imaging. Most of the light from sodium and mercury street lights is emitted in the green part of the spectrum, and exposures through a green filter will have a high background and a lower signal-to-noise ratio. To a lesser extent there will be light pollution leak through the red filter and very little through a blue filter, with a consequent skewed colour balance in the final image (see the Colour Gallery, p. 9). Most CCD cameras are particularly sensitive in the infrared part of the spectrum and least sensitive in the blue part. Typical exposure ratios through RGB filters are 1 : 2 : 4 (for an ST6), but will vary depending on the type of camera. To achieve a more realistic colour balance in a light-polluted sky a ratio of 2 : 6 : 4 may be more appropriate. A light pollution filter cannot be used in conjunction with colour filters.

The Art and Science of CCD Astronomy

Figure 9.6 The Flame Nebula in Orion (NGC2024) can be particularly elusive, but this 600 second image demonstrates the excellent contrast gain with a Lumicon broad-band light pollution filter. Meade 12 inch (30 cm) LX200 at f/3.3 on 28 September 1995.

Figure 9.7 The Veil Nebula in Cygnus (NGC6992). Again there is excellent contrast gain in this 1200 second image with a Lumicon broad-band light pollution filter. Meade 12 inch (30 cm) LX200 at f/3.3 on 2 September 1995.

Lunar and Planetary Imaging

Planetary images are not affected by light pollution, because they are bright, and exposures are consequently short. However, the effect on local seeing conditions of surrounding buildings will reduce high-resolution images, and the observer is advised to wait until the planet or the Moon is at its highest altitude before an attempt is made; again, it is best to wait until the early morning hours, when heat sources are at their lowest (Figures 9.8 and 9.9).

Conclusion

CCDs have revolutionised my interest in astronomy. I hope this chapter has given heart to city dwellers. I must add that I have recently moved to a country

Figure 9.8 Copernicus and surrounding area of the Moon taken with an ST6 at the prime focus of a Celestron 11 inch SCT. I used a lunar filter and 0.4 second exposure on 20 February 1994.

location with very dark skies where I hope to improve the quality of my images.

Figure 9.9 Three images of Jupiter during the Comet Shoemaker–Levy impact on 18 July 1994. I used an ST6 with my old Celestron 11 inch (28 cm) SCT.

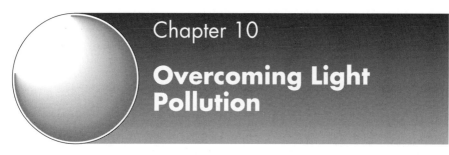

Chapter 10
Overcoming Light Pollution

David Ratledge

Introduction

I first began taking astrophotographs over thirty years ago, and although there was little light pollution then, film was slow and grainy. However, the more film improved over the years the worse light pollution got. It was very much one step forward, two steps backward! Where I live, on the northern fringe of the Greater Manchester conurbation, conventional astrophotography had become almost impossible. The arrival of CCDs, making the impossible possible, changed all that.

Setup

Like many others, I started into CCDs with the ST4 and its tiny chip. Despite its limitations this was a good trainer, and when the HiSIS camera, developed by the father of amateur CCDs, Christian Buil, appeared, I realised that here was a much more viable solution. My setup comprises a home-made 32 cm (12.5 inch) $f/6$ Newtonian mounted on a massive steel fork driven by a 25 cm (10 inch) Matthis worm gear. It is permanently mounted in a glass-fibre observatory, also home-made, at the bottom of my garden (Figure 10.1).

Figure 10.1
David Ratledge's home-made observatory, located at the bottom of his garden. A classical dome offers the best possible shielding from street lights.

For CCD imaging, in addition to the Newtonian I can piggy-back on it either a 15 cm (6 inch) $f/3.6$ Schmidt–Newtonian, or an 8 inch (20 cm) C8 Celestron tube assembly. The HiSIS (model 22) is based on the Kodak KAF0400 chip. I have found that the C8 with an $f/6.3$ telecompressor giving a focal length of about 1200 mm (50 inches) is ideal for both the pixel size and physical size of the Kodak chip. My method of combating light pollution is to take many 3 minute exposures and co-add them together. For operating the camera I use a (slow) 486 portable computer (Figure 10.2).

Image Taking Procedure

To explain my procedure for overcoming light pollution, I will describe how I took an image of the Horsehead Nebula. This had been an ambition of mine ever since my schooldays when I first saw it in a book

Overcoming Light Pollution 113

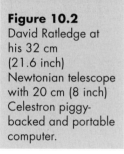

Figure 10.2
David Ratledge at his 32 cm (21.6 inch) Newtonian telescope with 20 cm (8 inch) Celestron piggy-backed and portable computer.

in the school library. However, from my northern location this was always going to be a challenge, as the Horsehead never rises high in the sky. My best photographic effort was made twelve years ago using the fastest colour film of the time (Ektachrome 400), but the result was hardly distinct. Since then light pollution had got much worse, but the CCD offered new hope.

For the Horsehead, the C8 with the telecompressor would be ideal for image scale, but because of severe light pollution at the Horsehead Nebula's low altitude it would be advisable to use a red filter (25A). Previous tests had shown that it would increase contrast. For objects higher in the sky this is unnecessary. The filter is mounted in a home-made filter wheel made for me by a colleague, Gerald Bramall, from my local society (Bolton), and can easily be swung into line when required (Figure 10.3).

Figure 10.3
Close-up of HiSIS CCD camera and home-made filter wheel.

My image taking procedure is to first align the C8 precisely with the 32 cm Newtonian, which will be used for finding the object and then guiding. The two telescopes are approximately aligned and firmly fastened together to try to eliminate any differential flexure between them. To align them precisely, a bright star near the object is selected, in this case, Betelgeuse. With a one second exposure and the camera operating in constant display in low-resolution mode, Betelgeuse is easily seen on the portable computer, with new images appearing every 5 seconds or so. Using the telescope drive hand control, the star's image can be brought to the centre of the chip. Then, to align the Newtonian with the C8, a centring device, invented and made by Gerald Bramall, is used (Figures 10.4 and 10.5). This enables the Newtonian's eyepiece to be moved bodily, in any direction across the focal plane, without moving the telescope. With Betelgeuse centred, both telescopes are exactly aligned.

The next step is to focus the CCD – with the red filter in place, of course. The method I use is that first described by Warren Offutt (*CCD Astronomy*, Winter 1995), namely diffraction focusing. This uses the fact that diffraction spikes on a bright star will be double unless the image is exactly in focus. The advantage of this method is that it is absolute – by looking at a single image you can determine whether it is in focus or not.

Overcoming Light Pollution

Figure 10.4 Centring device.

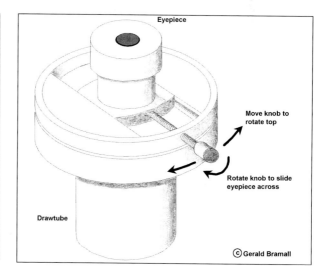

Figure 10.5 Diagram of centring device.

All others (to my knowledge) require a judgement as to whether a particular focus is better or worse than a previous one. Diffraction spikes are caused by spider vanes supporting the secondary mirror. However, with a C8 there are no spider vanes and therefore no diffraction spikes. So I have had to make some. I use an adjustable twin-bar arrangement that fits all my telescopes – again made by Gerald Bramall. The twin bars give stronger spikes than a single one would (Figure 10.6).

Using a continuous image display with a one second exposure and selecting the right display values (this is important), the diffraction spikes are big and clear. I

Figure 10.6
Adjustable twin bars for diffraction focusing on the C8.

find it best to orientate the bars so that the spikes will lie at 45° across the chip and are not aligned with pixel rows and/or columns. As soon as the spikes become single I know focus is achieved (Figure 10.7).

The next step is to find the object. To help with this I print finder charts using SkyMap and, in conjunction with setting circles, the Newtonian can usually be pointed at the object even when, as in this case, it is invisible. A quick, unguided 10 second exposure is enough to confirm that the object is there, and a quick touch on the drive controls centres it. The next task is to find a bright star for the Newtonian, which will be used for guiding. If you recall, the centring device allows the eyepiece to move in any direction. So, with the guiding eyepiece inserted, a quick turn of the knob and a suitable star can be centred effortlessly – usually about 10 seconds is all it takes!

We are now ready to expose. I usually take a dark frame first and have a cup of tea while it is running. As I said earlier, I've found that 3 minute exposures are about right for me and my light-polluted location. It

Figure 10.7
Appearance of a bright star with twin bars in front of telescope. Left: well out of focus. Centre: near to focus. Right: precise focus.

Overcoming Light Pollution

also lessens the chance of differential flexure of the two telescopes, and generally exposures do not suffer from "bleeding" on brighter stars. For the Horsehead, I will need at least ten exposures for a good signal-to-noise ratio on this faint object, more if possible. The first exposure is taken, guiding all the time. Guiding for 3 minutes is no hardship, especially compared with astrophotography, where 20 to 30 minutes was my norm! With the HiSIS there is no shutter as standard, so the telescope has to be uncovered at the beginning of the exposure and re-covered just before the end. This is simple to do and the operating software (QMips) beeps as a reminder at the appropriate time.

After two exposures, I take another dark frame and so on. I manage twelve lights and six darks in the end before the Horsehead disappears behind a neighbour's tree. One advantage of manually guiding is that for each exposure the image can be randomly shifted a few pixels. To do this the guide star is positioned in a different part of the reticule square of the guiding eyepiece (Figure 10.8). This means that the image will not fall on exactly the same pixels for every exposure, which mitigates against any chip defects.

Image Processing

Figure 10.8
Guide star positions.

At the end of the session the portable computer is returned to the house where, using the DOS utility

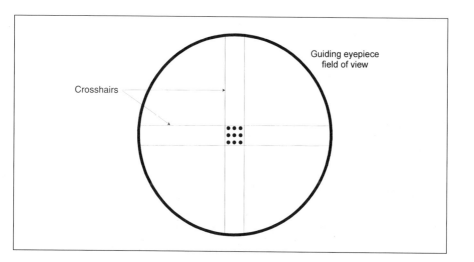

Interlink and a serial cable, the images are transferred to my desktop computer. The flat-field exposures, which are essential, are taken the following dusk. The camera will be left on the C8 in its exact position and focus until then. This means that if several images are to be taken in one night then they all have to be taken with the camera unmoved, in the same orientation. For the flats, I cover the front of the telescope with white paper and point it at the dusk sky. The paper eliminates any chance of imaging stars. I have also found clouds at dusk are equally satisfactory, which is fortunate for me, because if I had to wait for a clear evening then I could be waiting for weeks!

For calibration and processing I use the MiPS software developed by Christian Buil. I found it difficult to learn, but I would not swap it for any other now. The calibration and co-aligning of randomly shifted images is superb, with programs to automate each stage. The software also includes many exotic deconvolution techniques, but I generally find that for deep-sky objects, such as the Horsehead, simple log scaling is excellent. For final tweaking of the contrast I use PhotoStyler. For the Horsehead the colour balance was selected to match the red colour of the filter, resulting in a bright red nebula (see the Colour Gallery, p. 10).

For showing the images to societies I make 35 mm slides by shooting straight off the monitor. I do this in a darkened room, to cut reflections, with the camera set up on a sturdy tripod. I use a 105 mm lens and exposures are usually around 4 seconds at $f/11$ or $f/16$ with an ISO 100 film. It is important to check that the camera is square to the screen, otherwise the image will be distorted. Most monitors have a blue bias, so, where correct colour balance is required, the red needs boosting slightly.

So at last I have achieved my goal, an image of the Horsehead despite light pollution (Figure 10.9).

Other Images

Stephan's Quintet is an interesting interacting group of galaxies which has fuelled the argument over the meaning of redshifts. Because they are only around 13th magnitude, with the outer regions of the galaxies even fainter, a long (30 minutes total) exposure would

Overcoming Light Pollution

Figure 10.9   Horsehead Nebula.

Figure 10.10   Stephan's Quintet.

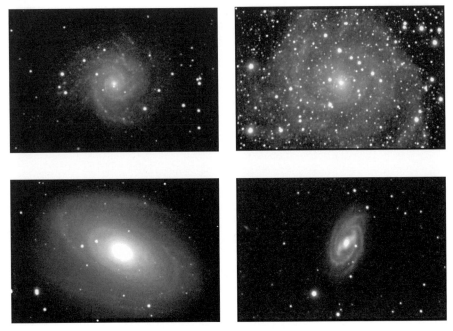

Figure 10.11
Clockwise from top left: M74, IC342, M109 and M81.

be needed. The galaxies were imaged over two separate nights. The images from the first night were simply added to those of a subsequent night (Figure 10.10). Another of the advantages of digital images!

Face-on galaxies, by their very nature, have low contrast and therefore are objects vulnerable to light pollution. As a result they had become almost impossible to photograph from my location. But they are some of the most attractive of targets for the imager. Fortunately, with my CCD and the procedure of co-adding many 3 minute exposures (9 in most cases), I have been able to record many of them in all their beauty. M74, M81, M109 and IC 342 are typical images (Figure 10.11), all taken with my C8 at $f/6.3$ and exposures of 27 to 30 min.

Conclusion

The advent of affordable CCD cameras for the amateur has rejuvenated my interest in Astronomy. Light pollution had made astrophotography, for all but the brightest objects, impossible from my location, but now even the faintest objects are in my range. Thanks to the CCD I have a lifetime of imaging to look forward to and I have even started building that bigger telescope I have always promised myself.

Chapter 11
Tri-Colour CCD Imaging

Nik Szymanek and Ian King

Introduction

Figure 11.1 Nik Szymanek in his observatory.

We first became acquainted, through a local astronomical society, in 1991. Both of us had recently and independently made first steps into CCD imaging with the purchase of an SBIG ST4 autoguider and Electrim

Figure 11.2 Ian King and his Brandon 4 inch refractor.

EDC1000TE CCD camera. Deciding to pool resources from then on, we purchased an SBIG ST6 and concentrated on producing some of the UK's first tricolour CCD images.

We worked from two observatories, equipped with a Meade 10 inch (25 cm) LX200 SCT and a Brandon 4 inch (10 cm) Apochromatic refractor. The next several years proved very productive, with the CCD proving a worthy competitor in the ongoing battle against light pollution.

Figure 11.3 Nik Szymanek and Ian King on La Palma.

A chance meeting with Konrad Malin-Smith in 1994 led to the first of many visits to La Palma, where the superb conditions allow the very best to be attained from amateur equipment.

Amateur Tri-colour CCD Imaging

Today's amateur astronomers routinely use colour emulsion to produce beautiful pictures of the night sky, but great care and effort is involved, because colour emulsion suffers many drawbacks, not least of all the dreaded reciprocity failure and colour shifts. Advanced amateurs have experimented successfully with the production of tri-colour astronomical photography, but this requires much skill in the darkroom. The success of tri-colour photography has peaked with the stunning work of David Malin, in which beautiful images containing a wealth of scientific data have been produced.

With the introduction of CCDs to the amateur astronomical community in the early 1990s it was natural that the principles developed for tri-colour photography were applied to electronic imaging. The pioneering efforts of Richard Berry, Jack Newton and Dennis di Cicco in America showed that amateur tri-colour CCD imaging was not only possible but far superior to emulsion astrophotography. The amazing sensitivity of today's CCD cameras means that even under light-polluted skies the amateur astronomer can produce images that are hardly inferior to those attained by professionals only a decade ago. The scientific worth of these images is surpassed only by their beauty, where stellar populations blaze with colour, where supernova remnants writhe with hydrogen-red filaments and glowing HII regions speckle star-forming spiral arms.

For the amateur to produce good results some fairly stringent hardware is required. As with all forms of astronomical imaging, a stable telescope with good optics and precise polar alignment is of the utmost importance. CCDs are ruthless in showing all forms of telescope deficiencies and, as we shall see later, many factors can combine to ruin a potentially good image.

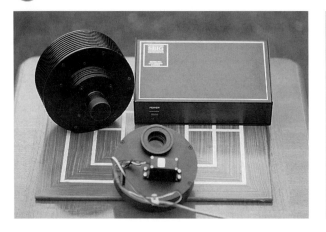

Figure 11.4 ST6 CCD camera and CFW6 motorised filter wheel.

For colour work it is necessary to use a filter-wheel or holder, preferably one that can be remotely controlled via computer, or where the filters can be changed without disturbing the position of the telescope (Figure 11.4). A combination flip-mirror–filter-holder can assist greatly in the acquisition of faint targets, and digital setting circles or telescope "GO TO" features are highly recommended, as are all forms of CCD autoguiders.

It is important to know the spectral sensitivity of the CCD camera being used for tri-colour imaging. Most CCDs are red-sensitive, with minimum sensitivity at blue wavelengths, and this has to be allowed for when taking filtered exposures. For example, the SBIG ST6 requires exposures with the ratio of 1 : 2 : 4 for red, green and blue respectively. The first range of CCDs featuring Kodak KAF0400 chips require eight times the exposure at blue wavelengths, amounting to very long exposure times overall. Focal reducers can help reduce exposure times and are useful if a wider area of sky is to be imaged, although when working under light-polluted skies the use of flat-fields is critical in order to reduce vignetting effects.

Two types of filter are available for tri-colour work. The Wratten type features a coloured gelatine dye as used in basic photography. These are cheap to purchase but extract a heavy toll on the incoming light because of heavy absorption. Far better are the interference or dichroic filters, which rely on fine coatings and the wave nature of light to produce very well-defined spectral bandpasses. These filters are far more forgiving in the amount of light absorbed when matched to the spectral response of the CCD, but can be quite expensive. Also required is an in-line infrared filter, which is important for reducing colour shifts produced by infrared excess from galaxies and high-star-content objects.

At the Telescope

The procedure for acquiring tri-colour images requires that the telescope is accurately focused through one of the filters (or a clear filter with the same physical thickness as the colour filters). The object of interest is centred on the chip and the exposure is started. The use of an autoguider is highly recommended, because it is important that the telescope is kept tracking accurately over the duration of the exposures so that when the individual components are co-added there is not too much drift between them. Once the exposure ratio has been decided (for example, 10 minutes red, 20 minutes green, 40 minutes blue) the first filtered components are acquired. It is not really necessary to take single long-duration exposures; a 40 minute exposure can be made up of four 10 minute components. This is far more forgiving on telescope drives, and offsets lost frames because of satellites, aircraft and other unforeseen eventualities.

In this way a sequence of red-, green- and blue-filtered monochrome images are taken, along with relevant support frames (dark and flat-field). The use of flat-field frames is recommended, and these must be matched to all filter configurations. It is highly instructive to view the filtered monochrome components of an object like the Crab Nebula. The red exposure highlights dramatically the red filamentary structure (see Figure 11.5), while recording little of the S-shaped background continuum, whereas the green and blue images register little of the filaments and record strongly the amorphous body of the nebula (Figures 11.6 and 11.7).

Similarly, M27, the Dumbbell Nebula, demonstrates a wealth of varying detail through each colour filter. The famous "apple-core" shape of the nebula, due to hydrogen emission, is easily imaged through the red filter. In stark contrast, the blue exposure, where light from oxygen and nitrogen dominates, captures a completely different view and shows faint horizontal

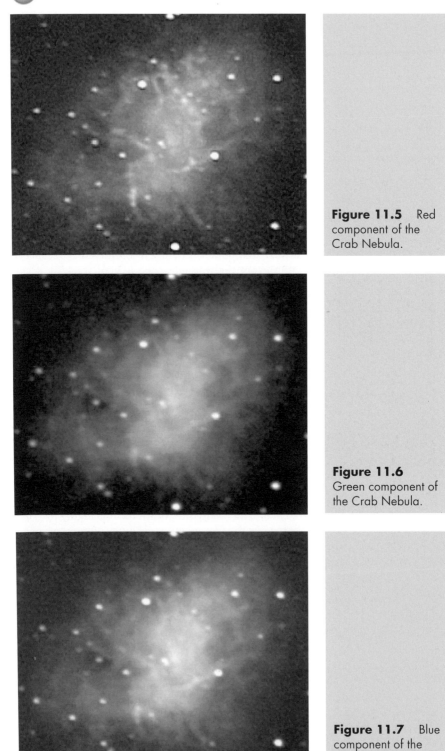

Figure 11.5 Red component of the Crab Nebula.

Figure 11.6 Green component of the Crab Nebula.

Figure 11.7 Blue component of the Crab Nebula.

Tri-Colour CCD Imaging

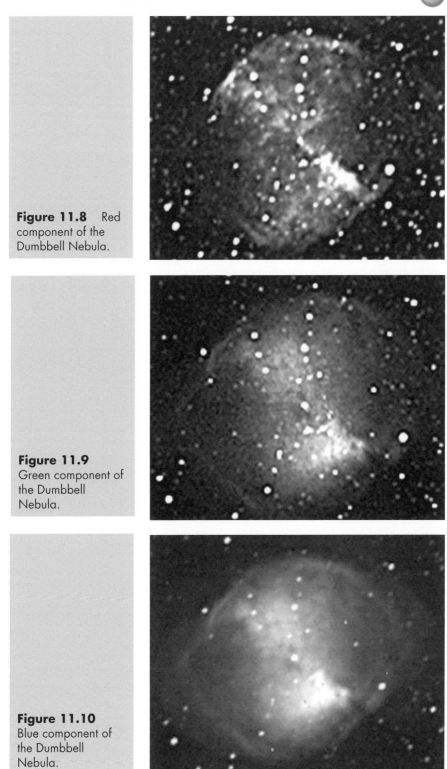

Figure 11.8 Red component of the Dumbbell Nebula.

Figure 11.9 Green component of the Dumbbell Nebula.

Figure 11.10 Blue component of the Dumbbell Nebula.

extensions (Figures 11.8 to 11.10). This selective recording of astronomical colours and structure occurs most dramatically on objects like M42 (see the Colour Gallery, pp. 000–000) or galaxies like M33 (see the Colour Gallery, pp. 000–000), where Population I and II stars are subtly imaged and red HII regions glow brilliantly.

Tri-colour Image Processing

Once the hard-earned images are safely stored on the computer hard disk, the task of image processing can begin. As with all CCD images the support frames are first applied. Flat-field frames are important for tri-colour work because the three coloured filters are likely to impart their own artefacts to the raw image, no matter how stringently they are cleaned or positioned. Once the images have been calibrated, the next step is to accurately co-align them. With a camera like the SBIG ST-6 each raw file is stored in 16 bit format (where each pixel can be displayed in any one of a shade of 65 535 grey levels). To combine the images into a 24 bit tri-colour file it is necessary to reduce each component to an 8 bit format, so the 16 bit raw image is first selectively scaled to maintain the required detail of the astronomical object contained in the image. Strictly speaking, each of the red, green and blue components should be scaled identically to maintain the correct colour balance.

There are two options at this stage. Co-alignment can be carried out manually, where, for example, the position of a common reference star can be noted using the numerical readout of the CCD camera software, and all images combined using that star. Alternatively, astronomical software such as SkyPro by Software Bisque will automatically combine the three images using a star centroid alignment routine. Once the three components are merged the 24 bit tri-colour file becomes a digital version of film emulsion. The beauty of digital colour files is that each of the red, green and blue frames is contained on its own "floating" level, where each component can be accessed for processing independently of the rest, allowing an unprecedented degree of manipulation.

Providing care has been taken to maintain the correct ratio of CCD camera spectral response and selective scaling, the tri-colour image should be an excellent true-colour representation of the astronomical object and to a degree that is far more accurate than any colour film. It is true that the sum of all exposures through the colour filters will probably exceed that of a single-emulsion exposure, but the CCD image will be far superior, particularly when light pollution probably wouldn't allow a long emulsion exposure anyway. For a selection of tri-colour images acquired by the authors see the Colour Gallery, pp. 11–13.

Tri-colour Processing Software

Many good programs are available to the CCD enthusiast, but one stands out above all others. This is Adobe PhotoShop, an industry-standard graphic manipulation package that is tailor-made for the processing of astronomical images. PhotoShop is a complete digital darkroom, offering an unprecedented amount of options, with no messy or dangerous chemicals to contend with. Here the images can be sharpened, blurred, lightened, darkened, resampled and, perhaps most impressive of all, *unsharp-masked* (a direct digital equivalent of the powerful techniques used by David Malin in tri-colour photography to bring forth a staggering amount of hidden detail), all at the click of a mouse button. The colour balancing of images is a highly subjective process, and it is all too easy to "add some colour here or there", so some restraint is needed. A good source of reference material is, once again, the tri-colour work of David Malin, which has been very carefully calibrated.

Excellent supplementary programs for image manipulation are also available. The Hidden Image by Sehgal Corporation allows the use of Maximum Entropy Deconvolution, a complex software routine that de-blurs an image, tightening star images and extracting hidden detail. The application of this sort of program requires that the individual monochrome components are first deconvolved before being co-added into a tri-colour file. This also applies to "log scaling", a brilliant scaling function which selectively brightens faint details without causing saturation of already bright pixels, and is tailor-made for large, dynamic-range objects like the Orion nebula.

Computer Hardware and Hard Copy

For the actual acquisition of tri-colour files a basic laptop computer is ideal. Most camera control software is fairly basic, and requires nothing more than a 386 computer. When it comes to image processing, however, the more powerful the computer the better. A 24-bit graphic card is necessary, as well as plenty of video and conventional RAM. The latest version of Adobe PhotoShop requires 16 megabytes of RAM and plenty of free hard-disk space. Complex mathematical programs like The Hidden Image require Pentium processors to run at a reasonable speed, and it is fair to say that all forms of image processing are greatly facilitated by high-specification computers. Lower-end systems can be used, but this will increase processing times dramatically.

Many options now exist for the production of hard copy of tri-colour images. High-quality 35 mm colour transparencies can be produced from images contained on floppy disks by graphic presentation companies, and many firms are now equipped with dye-sublimation printers for the production of high-quality prints. With the rapid improvement and cost reduction of standard inkjet printers the amateur astronomer can produce, cheaply and quickly, colour prints that are hardly inferior to standard colour photographs. Also, now that the digital age is truly here, it is possible to store hundreds of images on CD-ROM, cheaply and conveniently.

References

Di Cicco D (1993) The Universe in Colour. Sky & Telescope, May: pp. 34–40.

Malin D, Murdin P (1984) Colours of the Stars. Cambridge University Press.

Malin D (1993) A View of the Universe. Cambridge University Press.

Szymanek N, King I (1995) Going Mobile. CCD Astronomy, Summer: pp. 8–11.

Chapter 12

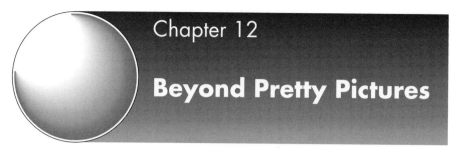

Beyond Pretty Pictures

George Sallit

Background

"At least ten years; certainly before I started work at Mount Wilson" – this was the response of the instrument technician when I asked about the last time film was used at the 100 in Mount Wilson telescope. Why have CCDs made such an impact on the professionals? As well as its high efficiency, the CCD is a very good measuring device. It is very linear and comes with built-in measuring frames. A similar revolution is now about to sweep amateur astronomy. Having taken some pretty pictures, some of you may want to carry out some science. Unlike others, I do not believe that you must, and if recreational astronomy is your interest, then carry on. There are too many distractions on our time without lecturing people on what should and should not be done. The best way to try this is to undertake a project that does not take all your imaging time; if you enjoy it then OK. So what science can be done? Well, the best areas to get started in are discovery and measurements. For discoveries the main areas are supernovae, asteroids and novae. For measurements the main areas are variable stars, novae, supernovae and active galactic nuclei.

Discovery

This is potentially the most exciting area, because you are trying to find something that has not been seen before. However, remember that there are many things that will appear on your images that will confuse you into believing you have found something when you have not. The important thing about discoveries is to check and double-check to make sure what you have found is indeed new.

Supernovae

With typical amateur telescopes and CCDs it is possible to record stars down to 16th or 17th magnitude with exposures of only 2 to 3 minutes. This will allow you to carry out supernova searches of many galaxies per night. In order to be successful at supernova hunting you need to image as many galaxies as possible. This requires a good telescope and drive, and if you use one with slewing such as the Meade LX200 then life gets easier. In fact there are some amateurs who use these scopes to carry out automatic supernova hunts. Having decided that you want to do this, you need to establish a library of reference images. There are three easy sources for amateurs: your own, catalogues, and on-line sources. For your own you will need to take a series of images that can be easily referenced. This is likely to take some time and effort. You should slowly build up a library and use other references in the meantime.

There are many catalogues available, but one of the best and cheapest is *Deep Space CCD Atlas: North* by John Rivers. This atlas contains galaxies imaged with typical amateur CCD equipment. It is difficult to get hold of, and I have included details at the end of the chapter. *The Carnegie Atlas of Galaxies* is also excellent, but be careful when comparing film and CCDs, because they have different spectral responses and a red star that is faint in photographs will be bright in red-sensitive CCDs.

Beyond Pretty Pictures

There are also many photographic catalogues, some of which are finding their way onto CD-ROMs. I find some of them have saturated cores, which makes comparison difficult – normally exactly where your "new" star is. There are many on-line sources, but one of the best is the digital scan of Palomar Observatory Sky Survey which is now available as an eight-CD-ROM set. Having got your references you must now choose which galaxies to image. Liller in his excellent book (see the references at the end of this chapter) analysed what type of galaxies supernovae are found in. His analysis showed they are not randomly distributed, but occur more frequently in spiral galaxies than ellipticals. Like Liller, I am not sure this is a real effect; rather, it could be caused by supernovae being more visible in spiral galaxies. He also found that some galaxies have more supernovae than others, the top ten being M83, M100, M101, M99, M61, NGC2276, NGC2841, NGC3814, NGC4157 and NGC6946.

Having decided on which galaxies you want to image, the next task is to actually image them. Remember, you are interested in discovering supernovae, not taking pretty pictures. Productivity is the name of the game. Having imaged the galaxies, you must accurately record all the necessary details such as telescope, camera, length of exposure, time, etc.. It is important that your time is precise to the nearest second. Please check your images as soon as possible, if not on the night of observation then next morning. When you compare your image to a reference there are going to be many false starts, red stars vs. blue stars on photographic plates, infrared (IR) objects that look fuzzy in CCDs, hot pixels, asteroids and satellites among others. One way to eliminate most of these is to take a second image 10 to 15 minutes later. If your supernova moves it isn't a supernova. However, it could be a new asteroid. Please check, check, check, and if you are sure its a new supernova then report it to Guy Hurst at *The Astronomer* and/or direct to IAU. Once found, you will need to measure its brightness and position.

Sometimes you may have to write off to experience an object that cannot be identified, as I did when I found a 13th magnitude "new" star on the edge of M64 (Figure 12.1).

Figure 12.1 A 600 second image of M64 using a 12 inch (30 cm) LX200 and ST7 with an f/3.3 focal reducer.

Asteroids

There are many thousands of asteroids out there waiting to be discovered. There are two main types, the main belt and the others, with the main belt congregating near the ecliptic. To improve your chances of finding the main belt, it is best to image between ±10° of the ecliptic, preferably a few hours east of the opposition point. This will allow you to follow any newly discovered asteroids over a period of a month. UK amateurs are at a disadvantage, because asteroids need at least three closely spaced observations to characterise their orbits. This means you need to have at least three clear nights in a sequence. It is preferable to have a series of five observations over a few weeks, with the more observations the better.

To increase your chances of finding a new asteroid you must image as large an area of sky as possible. With larger CCDs and new focal reducers this has become easier. I have found a good strategy is to take a wide field image with an exposure of 300 seconds, move the scope, take another image ensuring a small amount of overlap, and then repeat the sequence for a further five or six images. The images are then retaken starting at image 1. If any star has moved then you may have found an asteroid.

Some software has built-in blink comparators and it is possible to compare the images at the telescope while the camera is taking the next set of images. I find it best to include a reference object in the first image such as a galaxy or a known asteroid (Figures 12.2 and 12.3). This ensures that you know exactly where you are imaging. If you find an asteroid, check that it is not already known by using a good software package. If you do find one you must measure its position to sub-arc-second precision, know the time of the exposure to 1 second accuracy and estimate its brightness.

Figure 12.2 600 second image of NGC 772 taken on 27 September 1995. Unknown asteroid position is marked.

Figure 12.3 840 second image of NGC 772 taken 17 minutes later. Both this image and Figure 12.2 were taken with a 12 inch (30 cm) LX200/ST7.

Measurement

CCDs are excellent measuring devices. Because they are very linear they make the measurement of the brightness of objects (photometry) and the position of objects (astrometry) fairly straightforward. These measurements can be good to less than 0.05 magnitude and below an arc second, although great care is needed to achieve this type of accuracy.

There are two ways to measure the brightness of objects: differential and absolute photometry. Differential photometry compares the brightness of an unknown star to others in the same field of view, and absolute photometry uses standard stars to calibrate a system and then measure the brightness of an object by reference to these absolute standards. Differential photometry is the easiest approach for amateurs.

Any image must be properly calibrated using dark frames and flat fields; expect to take at least four dark frames and ten flat fields. The dark frames should be taken during the imaging session and a master created by averaging the original darks. This is particularly important when one is using non-temperature-regulated cameras. Taking a good flat field can be difficult, and care is needed to ensure that you do not add any more problems than you remove. I normally take an image of a white card in the observatory that is illuminated from a second card onto which I shine a standard white light. This double reflection gives a good even signal. An alternative is to put a sheet of white paper over the telescope aperture which acts as a diffuser and then directly illuminate a white screen. Because exposures are short there is no excuse for not taking at least ten images. These images are averaged to create the master flat field. Do NOT change the telescope–CCD system in any way, including the focus, between takes of the images and the flat fields.

Once the images have been calibrated, the brightness of the star can be measured. There are two main methods used to measure the signal from a star: aperture photometry and PSF photometry. Aperture photometry requires the total signal to be measured in an aperture that includes the star. From this signal the background contribution must be subtracted. This can be done by placing the measuring aperture over the star, recording the result, moving it to the background near to the star, recording that result and then taking one result

Beyond Pretty Pictures

from the other. Some software uses two apertures with the first measuring the signal from the star and the second, outer, one measuring the background. The program then automatically subtracts the background from the star signal and reports the value on a log scale. Remember that the magnitude scale is logarithmic.

The PSF method models the shape of the star using a point spread function. This function is found by measuring the average shapes of the stars in the image. This model is then used to calculate the total signal from the individual stars. This complex method is very useful when crowded star fields are being measured and aperture photometry is not possible.

The brightness of the star is now found by comparing its signal with that from the other stars in the field. The *Hubble Guide Star Catalogue* (*GSC*) gives access to over fifteen million stars whose magnitudes are reasonably well known. In fact it is unlikely that any image you take will not have at least three stars that appear in the *GSC*. It is important to use as many stars as possible, because the accuracy of these magnitudes can be between 0.1 and 0.3 from the correct value.

One other important consideration is filters. The spectral response of CCDs is different to previous measuring devices and they are generally sensitive in the red–infra red. To take account of this, measurements are taken through standard filters. Professionals have used many standards, but the most common is the BVR system.

Figure 12.4 BVR system divides the spectrum into three broad regions.

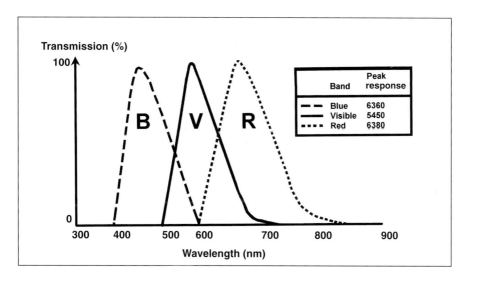

This divides the spectrum into three: blue, visible and red. Special filters are used that transmit light according to these curves. Measurements are made, through these filters, of standard stars and the results allow calibration of the telescope–CCD system. Using these calibration factors allows subsequent measurements to be corrected to the BVR system and compared to measurements made by other observers. While I accept that this is the right way to measure variable stars, good work can be done without filters. If you enjoy doing this type of work then try using just a V filter. If you then wish to continue there are a number of companies that make filter wheels and photometric filters. I can particularly recommend the one produced by Norman Walker.

The measurement of the position of objects (astrometry) also requires good calibrated images. It is important to have as many known stars as possible: 3 stars are the minimum to make a measurement, 5 the practical minimum and over 8 a bonus. One of the best references for star positions is again the *Hubble GSC*, which is reasonably accurate over small areas.

The first step in the astrometry process is to match the CCD image with stars from the *Hubble GSC*. This is more difficult than it sounds, as CCD images generally show more stars than the *Hubble GSC*. From the *Hubble GSC* the right ascension and declination of the reference stars are found, and their XY positions measured on the CCD image. A polynomial equation is then derived that relates the right ascension and declination of the stars to their XY positions. Using this equation the right ascension and declination of any object on the CCD can be calculated. With some software packages an estimate of the errors can also be calculated. As with photometry, I suggest you get a specialised software package that automates as much of the process as possible.

Software

There is a bewildering array of software on the market, and it is important you get the right package because it will save a great deal of time and effort. You will need two types of software, one to produce star charts and the other to measure the position and brightness of the object. I confine myself here to PC-based software because I have no experience of Mac, Archimedes or others. For star maps, telescope control packages, asteroid and comet ephemerides I recommend Guide (DOS-based), Sky and Megastar. They produce good star maps, calculate the positions of thousands of known asteroids and give the GSC magnitudes and positions of stars down to 15th magnitude.

For measuring the positions of objects you must use a purpose-designed program, as high precision is needed. I have an aversion to the "old-fashioned" programs. We no longer have to put up with programs that have an interface of a flashing cursor. If you have to measure the position of 10 objects using 20 stars per measurement then the sheer frustration experienced with this type of interface is enormous. I particularly recommend MIRA, WinMiPS and CCDAST, with CCDAST getting my overall vote. Herbert Raab's Astrometrica program is a specialised astrometry program and is highly recommended, but I have no first-hand experience of it.

For photometry I recommend MIRA, WinMiPS and CCDIR. There are also some good professional packages now available, but some of the interfaces are "clunky".

In summary, I suggest that you get the right software for your particular interest and try to gain experience of using it. Try following known asteroids or variables, and after a few weeks you should have ironed out all the wrinkles. Then try measuring something new. Be patient, and the rewards will come; but remember that this is your hobby and if it gets too much of a chore then you have got the wrong software or the wrong interest!

References

Reference books to give you more details

Astronomical Society of the Pacific Conference series vol. 23.

Buil C, CCD Astronomy – Construction and Use of an Astronomical CCD Camera. Willmann Bell, Richmond, Virginia 23235, USA.

Howell S (ed) Astronomical CCD Observing and Reduction Techniques. Astronomical Society of the Pacific Conference series, vol. 23, ISBN 0-937707-42-4.

Liller W, The Cambridge Guide to Astronomical Discovery. Cambridge University Press.

Haitchuck R, Henden A, Truax R (1994) Photometry in the Digital Age. CCD Astronomy, Fall, pp. 20–23.
Bessel M (1995) UBVRI Filters for Photometry. CCD Astronomy, Fall, pp. 20–23.
Di Cicco D (1996) Hunting Asteroids. CCD Astronomy, Spring.

Articles from magazines that are helpful

CCD Astronomy is an excellent magazine and easy to read, as is *Observatory Techniques* from Mike Otis, 1710 SE 16th Avenue, Aberdeen, SD 57401- 7836, USA.

GUIDE. Project Pluto. Ridge Road. Box 1607, Bourdoinham, ME 04008, USA.
THE SKY. Software Bisque, 912 Twelfth Street, Suite A, Golden, CO, 80401, USA.
MEGASTAR. E.L.B Software 8910 Willow Meadow Dr., Houston, TX 77031-1828, USA.
WinMiPS/MIPS WinMiPS is freeware. MIPS. In the USA: RX Design, 1753 Elmwood Dr., Rock Hill, SC 29730. In the UK: True Technology. Woodpecker Cottage, Red Lane, Aldermaston, Berkshire RG7 4PA.
MIRA. 1304 East Eighth Street, Tuscon, AZ 54719, USA.
CCDAST. Computer Aided Astronomy, PO Box 1814, Camarillo, CA 93011-1814, USA.
ASTROMETRICA. Herbert Raab, Schrammlstrasse 8, A-4050 Traun, Austria.
CCDIR. Unified Software Systems, PO Box 23875, Flagstaff, AZ 86002-3875, USA.

Good software packages

Appendix A
CCD Camera Manufacturers

Many of the cameras listed below are available through dealers or importers rather than direct from the manufacturer.

Manufacturer's name and postal address	Model name	Web pages or email address
Santa Barbara Instrument Group PO Box 50437, 1482 East Valley Road #33, Santa Barbara, CA 93150, USA	SBIG ST	sbig@sbig.com
LE21M 188 Chemin du Bois Don, BP 38, 69380 Lozanne, France	HiSIS*	
Starlight Xpress FDE Ltd, Bodalair house, Sandford Lane, Hurst, Berkshire RG10 0SU, UK	SX	http://www.demon.co.uk/astronomer
Axiom Research Inc. 3340 N. Country Club, Ste. 103, Tucson, AZ 85716, USA	AX, Viper	http://www.axres.com/axiom/index.html
SpectraSource Instruments 31324 Via Colinas, Suite 114, Westlake Village, CA 91362, USA	Lynxx, Teleris	spectrasrc@delphi.com
Electrim Corporation 356 Wall Street, Princeton, NJ 08540, USA	EDC	
Helius Designs The White House, Aldington, Evesham, Worcestershire WR11 5UB, UK	Io	100674.431@compuserve.com
Sirius Instruments 141 N. Charles Avenue, Villa Park, IL 60181, USA	CWIP	http://rampages.onramp.net/~tennant/sirius.htm
Meade Instruments Corporation 16542 Millikan Avenue, Irvine, CA 92714, USA	Pictor	
IISIS Inc. 3463 State Street, Suite 431, Santa Barbara, CA 93105, USA	Compuscope	info@compuscope.com

*The HiSIS22 is also made in the USA by RXDesign, 1753 Elmwood Drive, Rock Hill, SC 29730, USA.

Manufacturer's name and postal address	Model name	Web pages or email address
Celestron International 2835 Columbia Street, Torrance, CA 90503, USA	PIXCEL	
Apogee Instruments Inc. c/o Pixel, 6114 LaSalle Avenue, Suite 503, Oakland, CA 94611, USA	AM	http://www.apogee-ccd.com
Make your own See Appendix E – *The CCD Camera Cookbook*	Cookbook	

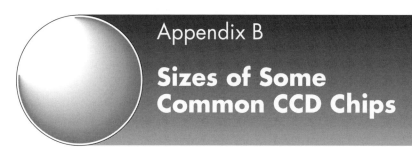

Appendix B
Sizes of Some Common CCD Chips

Manufacturer	Name	Number of pixels		Size of pixel		Array size	
		Width	Height	Width μm	Height μm	Width mm	Height mm
Texas Instruments	TC211	192	165	13.75	16	2.64	2.64
	TC255	336	243	10	10	3.3	2.4
	TC241	375	242	23	27	8.6	6.5
Kodak	KAF-0400	768	512	9	9	6.9	4.6
	KAF-1600	1536	1024	9	9	13.8	9.2
Thomson	TH7895M	512	512	19	19	9.7	9.7
Tektronix	TK512	512	512	27	27	13.8	13.8
	TK1024	1024	1024	24	24	24.6	24.6
Philips	FT12	512	512	15	15	7.68	7.68

Appendix C
CCD Field of View for Various Focal Lengths

Manufacturer	CCD	Field of view in minutes of arc (1° = 60 min of arc)							
		300 mm FL		600 mm FL		1200 mm FL		2000 mm FL	
		Width	Height	Width	Height	Width	Height	Width	Height
Texas Instruments	TC211	30	30	15	15	8	8	5	5
	TC255	38	28	19	14	9	7	6	4
	TC241	99	74	49	37	25	19	15	11
Kodak	KAF-0400	79	53	40	26	20	13	12	8
	KAF-1600	158	105	79	53	40	26	24	16
Thomson	TH7895M	111	111	56	56	28	28	17	17
Tektronix	TK512	158	158	79	79	40	40	24	24
	TK1024	282	282	141	141	70	70	42	42
Philips	FT12	88	88	44	44	22	22	13	13

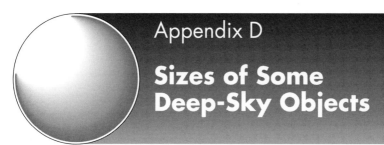

Appendix D
Sizes of Some Deep-Sky Objects

Common name/description	Catalogue number	Size arc min
Crab Nebula	M1	6×4
Dumbbell Nebula	M27	8×6
Andromeda Galaxy	M31	180×40
Hercules Globular	M13	17×17
Orion Nebula	M42	70×60
Whirlpool Galaxy	M51	9×7
Ring Nebula	M57	1.4×1
Galaxy in Leo	M65	9×2
Owl Nebula	M97	3×3
Face-on galaxy	M101	28×28
Edge-on galaxy	NGC4565	15×2

Appendix E
Bibliography

Title	Author	First Published	Publisher	Internet
CCD Astronomy	Buil C	1991 (English edition)	Willmann-Bell	
Introduction to Astronomical Image Processing	Berry R	1991	Willmann-Bell	
Choosing and Using a CCD Camera	Berry R	1992	Willmann-Bell	
The CCD Camera Cookbook	Berry R, Kanto V, Munger J	1994	Willmann-Bell	
CCD Astronomy magazine (quarterly)		1994	Sky Publishing	http://www.skypub.com

Appendix F
Astronomical Image Processing Software

Software Name	Author, distributor	Email
AstroIP	Richard Berry, Willmann-Bell	
Epoch2000ip	Meade Intruments Corp.	
Hidden Image	Sehgal Corp.	sehgal@achilles.net
Imagine32	ISIS, Compuscope	info@compuscope.com
MiPS, WinMiPS	Christian Buil, LE21M	
Mira	Axiom Research	info@axres.com
SkyPro	Software Bisque	
SuperFix	Bruce Johnston Computing	Bjohns7764@gnn.com
CCDIR (photometry)	Unified Software Systems	

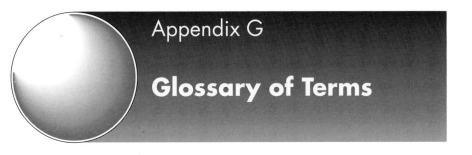

Appendix G

Glossary of Terms

A/D converter	Analogue-to-digital converter, used for the conversion of electrons produced by the incidence of photons on the CCD into a digital (binary) number.
bias frame	An exposure of zero time with the CCD covered. It is an image of the electronic noise present in the camera.
binning	Electronically joining pixels together to make larger ones. Larger pixels have less noise, are faster, and have a larger capacity but less resolution.
bleeding/blooming	Ugly (to some) spikes produced when a pixel saturates and the signal spills into the rest of the column of pixels. Use shorter exposures to avoid. Some chips have anti-blooming gates or drains, which mitigate the effect but are less sensitive as a result
calibration	Correcting the image for both dark current and non-uniformity.
CCD	Charge-coupled device.
dark frame	An image produced by covering the CCD and taking a blank exposure! The exposure time is normally the same length as that of the exposed (or light) frame.
deconvolution	Restoration of the image to what the image (probably) would have been had there been no atmospheric turbulence, poor guiding, poor focusing or imperfect optics etc. to degrade it. Popular deconvolution algorithms are Maximum Entropy, Wiener Filtering, Lucy-Richardson and Van Cittert.

Glossary of Terms

FITS	Standard file format favoured by professional astronomers. Supported by some of the better image processing software. Produces large (uncompressed) files.
flat field	An image of a totally uniform object such as a white screen. Used to correct the image for non-uniformity caused by the CCD itself and the optics.
GIF	File format which produces compressed 256 colour files. Favoured by the Internet. Can hold text in a comment block. Some dispute over whether it is proprietary or not.
JPEG, JPG	File format which produces small compressed files but involves some loss of information.
light frame	The opposite of a dark frame – an exposure taken with the CCD uncovered, i.e. it will record the object.
noise	The unwanted part of the signal. Caused not only by deficiencies in the CCD and electronics but also by the fundamental properties of light itself. Can be minimised but not eliminated.
photon	Although light behaves as a wave it also behaves as a particle called a photon; a photon is the smallest quantum of light.
pixel	Short for picture element. In our case either the light-sensitive cell on the CCD chip or the smallest display resolution on a monitor.
raw image	The image before any calibration or correction has taken place.
thermal frame	An image of the thermal (heat) signal present during an exposure. It increases in proportion to exposure length and temperature. Produced by subtracting a bias frame from a dark frame.
TIF, TIFF	Standard file format used by graphics professionals and publishers. Can be compressed or uncompressed and hold "true" colour.
unsharp masking	The former darkroom technique brought into the digital age. It enhances the contrast of fine details, by creating a blurred copy of the image, which is then subtracted from the original. Usually works best for planets but is also good for globular clusters.

Glossary of Terms

well capacity   How many electrons each pixel can hold before saturation. At saturation the linear response ends and bleeding can occur.

Contributors

David Ratledge is a Civil Engineer working in local government, formerly specialising in structural design but more recently in I.T. and graphics/publicity. He lives in Adlington, Lancashire, UK with his wife Julie and son John. He is chairman of Bolton Astronomical Society.
His E-mail address is: 100632.2746@compuserve.com
Web pages http://ourworld.compuserve.com/home pages/david_ratledge

Dave Petherick is an automotive research engineer living in Tottenham, Ontario, with his wife, Brenda, fifteen cats, two computers, one CCD camera and, of course, his just-barely-portable home-built 8-inch (20 cm) reflecting telescope. He is an active member of the South Simcoe Amateur Astronomers.
E-mail: 73627.226@compuserve.com

Brian Colville is an agricultural specialist with a local conservation authority and lives with his wife, Sandy, in Sunderland, a town 60 miles (100 km) north-east of Toronto, Ontario, Canada. He survived a car accident during production of this book – fortunately he could still type!
E-mail: 103406.1220@compuserve.com

Gregory A. Terrance lives near the small town of Lima, New York, with his wife Dawn and three children. He works for an automotive firm as a technical photographer and desktop publisher.
E-mail: gregoryt@frontiernet.net
Web pages: http://www.frontiernet/~gregoryt/

Tim Puckett is a 34-year-old small business owner from Villa Rica, Georgia, USA. Tim's work has appeared in many publications and on several US television channels including "Good Morning America".
E-mail: tpuckett@mindspring.com

David Strange is a farmer on the South Dorset Heritage Coast in the UK, and is married with three daughters. A past chairman of Wessex Astronomical Society, he also gained the 1994 Dorset Archaeological Award for his involvement in research on an iron age site.
E-mail: 100614.3525@compuserve.com

Luc Vanhoeck is a laboratory manager for a large pharmaceutical multinational and is married with three daughters. He has been an active astronomer in Belgium for over twenty years, playing a leading role in astronomy camps, magazines, astronomical exhibitions, lectures, etc.
E-mail: vanhoeck@vvs.innet.be
Web pages: http://ourworld.compuserve.com/home pages/Luc_Vanhoeck

John Sanford is a 57-year-old college professor teaching photography (including astrophotography) and lives in Southern California. He is president of Orange County Astronomers, a 605-member astronomy club, and received the G. Bruce Blair medal of Western Amateur Astronomers in 1983.
E-mail: jonsanf@interserv.com
or 100410.464@compuserve.com

Adrian Caterall is a consultant gastroenterologist and physician living in Hertfordshire (formerly Dulwich, South London) with his girlfriend.
E-mail: 100120.351@compuserve.com

Nik Szymanek is a train driver on the London Underground and **Ian King** is an insurance underwriter from Essex, England. They are both members of Havering Astronomical Society where they met in 1991 and decided to pool resources for CCD imaging.
E-mail: 100304.2143@compuserve.com (Nik Szymanek); 100746.3521@compuserve.com (Ian King).
Web pages: http://ourworld.compuserve.com/home pages/Nik_Szymanek

George Sallit is head of radiation safety at the UK's Atomic Weapons Establishment and is happily married to Jennifer. He is a member of both Reading and Newbury Astronomical Societies.
E-mail: 100573.2667@compuserve.com

Colour Gallery

Some of the best images from the Contributors

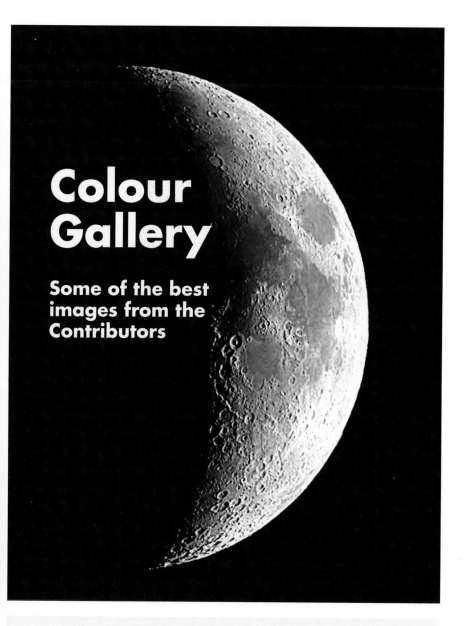

A 25 image Moon mosaic by Gregory A. Terrance with C5 and Lynxx CCD Camera

Dave Petherick

Images clockwise from top: nucleus of Comet Hyakutake, M82 galaxy in Ursa Major, M13 globular cluster in Hercules. All with 8 inch (20 cm) Newtonian and Cookbook CCD camera. False colour.

Colour Gallery
Brian Colville

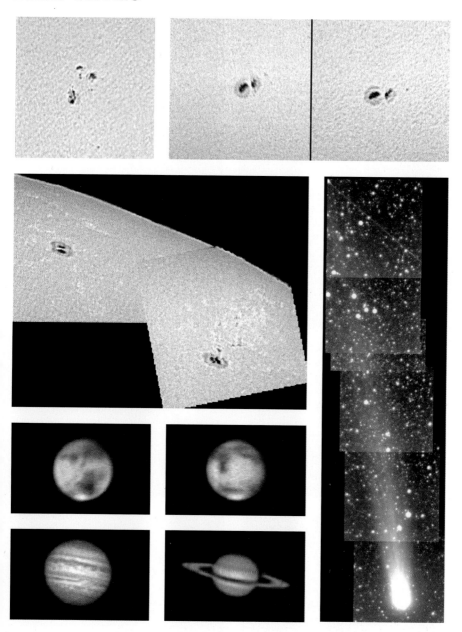

Images clockwise from top: solar regions, solar regions pair twenty-four hours apart. Comet Hyakutake on 29 March 1996 with 50 mm lens and 2 minute exposures. Saturn on 15 October 1994 at f/20. Jupiter on 23 June 1995 at f/20. Mars in February/March 1995 with Wratten No. 25 filter at f40, solar mosaic exposure 1/100 second. All with 25 cm (10 inch) SCT, Lynxx CCD camera and false colour unless stated otherwise.

Gregory A. Terrance

Images clockwise from top left: M51 (16 inch (40 cm) f/5 and exposures R:G:B, 7:10:30 minutes – ST6), M16 the Eagle nebula (8 inch (20 cm) f/5 Newtonian and ST6), M27 the Dumbbell Nebula (16 inch (40 cm) f/5 – ST6), M17 the Omega Nebula (8 inch (40 cm) f/5 Newtonian – ST6), Comet Hyakutake (5 inch (12.5 cm) SCT at f/6.3 – 30:30:30 seconds with ST6) and the Horsehead Nebula (8 inch (20 cm) f/5 Newtonian – ST6).

Colour Gallery
Tim Puckett

Images clockwise from top: Comet Hyakutake taken on 11 April 1996 (false colour), Tim Puckett at work, Omega Centauri taken on 4 March 1995 (tri-colour), Comet P/Schwassmann3 taken on 14 March 1995 (false colour), Comet Hale–Bopp taken on 24 April 1996 (false colour), M42 Orion Nebula taken on 2 March 1995 (tri-colour). All using 12 inch (30 cm) SCT and ST6 CCD camera.

David Strange

Colour Gallery

C/1996 B2 0101h U.T. March 17th 1996

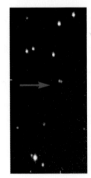

Images clockwise from top left: Comet Hyakutake (135 mm lens and SX CCD camera). The gravitationally lensed Quasar, and transits of Ganymede and Io across Jupiter, both with 500 mm (20 inch) f/4 Newtonian and SX. False colour.

Colour Gallery
Luc Vanhoeck

Images clockwise from top: the Helix Nebula, the flame nebula with the Horsehead Nebula, the Lagoon Nebula, and the Trifid Nebula. All with 4 inch (10 cm) f/5 Genesis refractor and ST6 camera. Tri-colour exposures of R:G:B, 10:10:10 minutes. Taken in collaboration with Staf Geens, Tim Polfliet and Rik Blondeel.

John Sanford

Colour Gallery

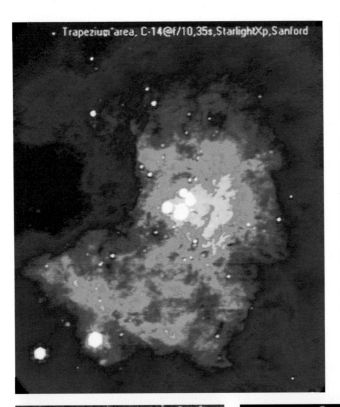

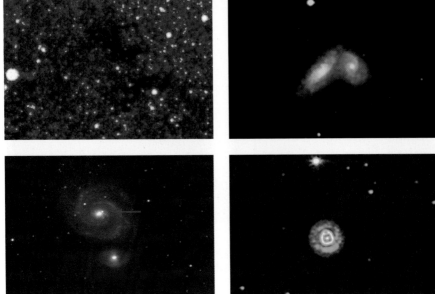

Images clockwise from top: the central region of M42 (C-14), the Siamese Twins in Virgo (C-14), the Eskimo Nebula in Gemini (C-14), supernova SN1994i in M51 imaged immediately after visual discovery in collaboration with Wayne Johnson and Doug Millar (4 inch (10 cm) f/10 refractor) and Barnard 143 dark nebula in Altair (135 mm telephoto lens). All with Starlight Xpress CCD camera. False colour.

Colour Gallery
Adrian Catterall

Images clockwise from top: Comet Hyakutake. 12 inch (30 cm) SCT at f/3.3 with ST7 in 9 μm pixel mode, exposures R:G:B, 60:100:200 seconds. Taken from South London: Veil Nebula, 12 inch (30 cm) SCT at f/7.5 with ST7 in 18 μm pixel mode, exposures R:G:B, 10:20:40 minutes; M13 in Hercules. Meade 12 inch (30 cm) SCT at f/7.5 with an ST7 in 18 μm pixel mode, exposures R:G:B, 5:5:10 minutes; M82 in Ursa Major. Meade 12 inch SCT at f/10 with an ST6, exposures, R:G:B, 80:180:120 seconds. Total eclipse of the Moon, April 3–4th 1996 from South London. 70 mm (2.8 inch) f/6 refractor and ST6 with CFW6 colour filter wheel. exposures R:G:B, 0.5:0.5:1.0 seconds.

David Ratledge

Images clockwise from top: Horsehead Nebula (exposure 36 minutes No. 25 red filter). M15 globular cluster (15 minutes), galaxies NGC925 and NGC891 (both 27 minutes). All taken with a C8 at f/6.3 and HiSIS CCD camera. False colour.

Colour Gallery
Nik Szymanek and Ian King

Images clockwise from top: the Orion Nebula M42 (exposures R:G:B, 14:12:20 minutes), the Horsehead Nebula (14:10:10 minutes), NGC1977 and M33 (both 10:10:20 minutes). All taken with Brandon 4 inch (10 cm) f/7 refractor and ST6 from La Palma Tri-colour.

Nik Szymanek and Ian King

Images clockwise from top: the Lagoon Nebula (4 inch (10 cm) refractor/ST6 – exposures R:G:B, 10:10:20 minutes), the Trifid Nebula (10 inch (25 cm) SCT/ST6 – 20:20:40 minutes), the Crab Nebula (10 inch (25 cm SCT/ST6 – 40:40:60 minutes) and NGC253 (10 inch (25 cm) SCT/HiSIS22 – 20:20:40 minutes). All from La Palma except the Crab from the UK. Tri-colour.

Colour Gallery

Nik Szymanek and Ian King

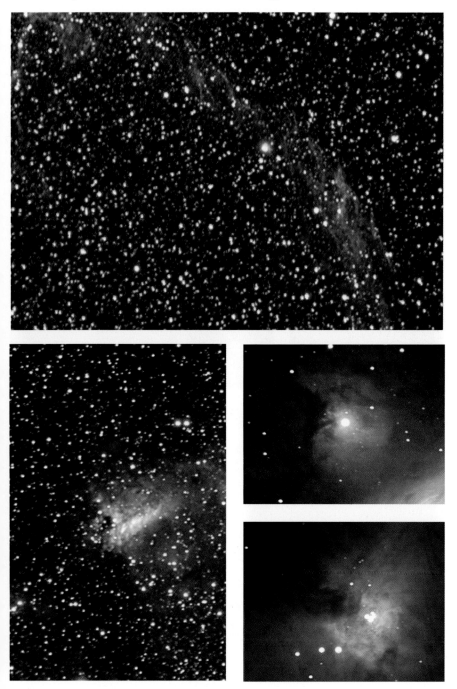

Images clockwise from top: the Veil Nebula (4 inch (25 cm) refractor), M43 (10 inch (25 cm) SCT), the Trapezium (10 (25 cm) inch SCT) and M17 (4 inch (10 cm) refractor). M43 and the Trapezium taken from the UK, the Veil and M17 from La Palma. All taken with ST6 and Tri-colour. Exposures R:G:B, 10:10:20 minutes for Veil and M17, 5:5:5 minutes for M43 and Trapezium.

George Sallit

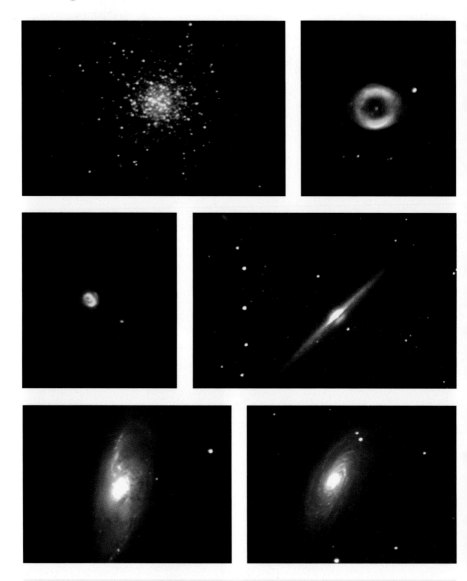

Images top left to right: M13 globular cluster (tri-colour exposures at f/6.3 R:G:B, 40:40:160 seconds and M57 the Ring Nebula (tri-colour exposures at f/10 R:G:B, 320:320:4x320 seconds) both with 12 inch (30 cm) SCT and Starlight Xpress CCD camera.
Middle left to right: NGC 7662 planetary nebula (exposure at f/6.3 80 seconds) and NGC 4565 (exposure at f/3.3 500 seconds) both with 12 inch (30 cm) SCT and SBIG ST7 CCD camera. False colour.
Bottom left to right: M106 (exposure at f/6.3 7x200 seconds and M88 (exposure at f/6.3 7x200 seconds) both with 12 inch (30 cm) SCT and HiSIS CCD camera. False colour.

Colour Gallery
Miscellany

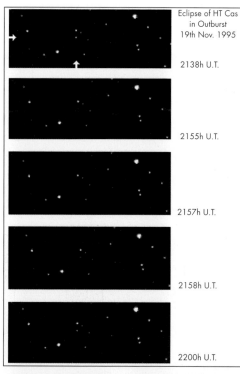

Eclipse of HT Cas in Outburst 19th Nov. 1995

2138h U.T.

2155h U.T.

2157h U.T.

2158h U.T.

2200h U.T.

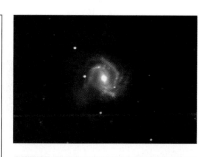

Images clockwise from top left:
HT Cassiopeia sequence by David Strange with 50 cm (20 inch) f/4 Newtonian and Starlight Xpress. M61 by George Sallit with 30 cm (12 inch) SCT and HiSIS22. Archimedes region mosaic by Gregory A. Terrance 40 cm (16 inch) Newtonian and Lynxx CCD, NGC 7331 by David Ratledge with C8 and HiSIS CCD, Saturn and Jupiter tri-colour by Adrian Catterall with 30 cm (12 in) SCT and ST6 CCD.

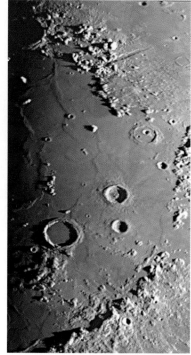

Colour Gallery

Tailpiece – the ultimate mosaic

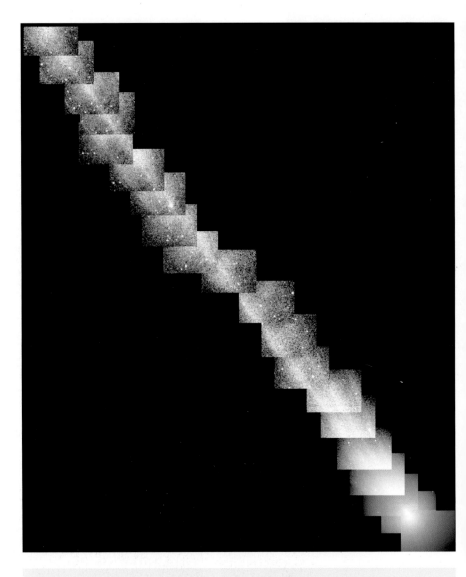

Comet Hyakutake by Brain Colville and Dave Petherick. 15 second exposures with 10 inch (25 cm) at f/4.6 and CB245 CCD camera.